THE BRITISH COAL-MINING INDUSTRY
retrospect and prospect

The British Coalmining Industry

retrospect and prospect

A. R. Griffin

[ALAN]

MOORLAND
PUBLISHING

To my sons

ISBN 0 903485 41 9

© A. R. Griffin 1977

COPYRIGHT NOTICE

Printed in Great Britain by
Wood Mitchell & Co Ltd, Stoke-on-Trent

For the Publishers
Moorland Publishing Company
The Market Place, Hartington,
Buxton, Derbys, SK17 0AL.

Contents

Plates

Figures

Tables

Preface

Coalmining has been well served by historians. The work of Professor J. U. Nef, Ashton and Sykes for the eighteenth century and Robin Page Arnot on mining trade unionism are too well known to need any detailed mention here. R. L. Galloway's two-volume *Annals of Coal Mining and the Coal Trade,* written originally as a series of articles for the *Colliery Guardian,* is not sufficiently well known or appreciated however. Galloway was a distinguished mining engineer, and he is one of the few mining historians whose knowledge is matched by his understanding of the subject.

In recent years, a number of regional and local historians have made valuable contributions to the study of mining history, Baron Duckham being perhaps the outstanding example; although I also like the studies of Donald Anderson who has Galloway's combination of scholarship and practical understanding. Books on particular aspects of mining — technological, economic, commercial, social and so on — have appeared in profusion in recent years. What has been lacking is a reasonably comprehensive and concise general history covering Great Britain, and the present volume has been designed to fill that need.

Of course, there is a limit to what can be compressed into 90,000 words or so, and I have had to be selective. What I have attempted to do is to construct a general framework, and within it to concentrate on those topics which I regard as particularly important or where I have something new to say. Inevitably, since my own research and that of my son, Dr Colin Griffin, has been mainly concerned with the East Midlands, more examples have been drawn from that region than its relative importance warrants; but I have tried to avoid allowing regional bias to distort the main thrust of the argument.

In previous books I have acknowleged the help of various friends who have supplied information or helped in some other way. I thank them again, but would wish particularly to mention my son Ian Karl Griffin for his artwork, Mr E. A. Dyson of Mansfield whose pioneer field-work on an opencast site has provided excellent material for illustration and Mr Herbert Booth whose phenomenal memory of the mining world of the late nineteenth and early twentieth century has been an invaluable part of my education. My thanks are also due to the National Coal Board for invaluable financial help. May I make it clear however that the views expressed in this book are entirely my own.

A. R. Griffin
Department of Economic History,
University of Nottingham

November 1976

Introduction

Britain's coal seams were laid down in the carboniferous period, some 250 million years or so ago. Much of Britain was then part of a low-lying subtropical swamp with rivers flowing into it and the whole area was gradually, though not continuously, sinking. Trees and other plants grew profusely, and as they died they were 'deposited as waterlogged masses of vegetable debris'.[1] A similar process can be seen today in the Ganges Delta which is carpeted by layers of decayed vegetable matter (commonly called peat).

Periodically, an increased rate of sinking would result in a layer of peat being completely covered by mud deposited by the rivers flowing into the swamp. This process, where alternate layers of peat and mud were laid down, continued throughout the carboniferous period. In later geological periods, the coal measures had rocks of various kinds superimposed on them, and being subject to increasing pressure and temperature, the peat changed into coal and the mud into rock of one sort or another, sandstone and shale being the most common. Fossil leaves, plant stems and tree roots are often found in coal, and its vegetable origin can sometimes be detected also by particles of fibre. This is particularly true of the brown coal (or lignite) found in many countries, but some types of bituminous coal also contain fibrous particles.[2]

Coals are often ranked according to the proportion of volatile matter and carbon they contain. Lignite is the most volatile, having about 67 per cent pure carbon, while anthracite, a hard coal found mostly in South Wales and used for steam raising, is the least volatile, having about 93 per cent pure carbon. Anthracite burns almost smokelessly. Bituminous coals contain varying proportions of carbon, 88 per cent being a fair average. Cannel, a dense shiny coal which has about 84 per cent carbon, burns with a flame like a candle, and used to be in great demand for making gas and by-products because of its volatility.

If we regard peat as being at one extreme of a geological time scale, then anthracite is at the other. However, we cannot assume that the older the seam, the deeper it will lie in the ground, because earth movements — some violent and some gradual — have distorted the earth's crust. Indeed, some coal seams, thrust to the surface millions of years ago, have suffered erosion by the action of sun and rain.

A detailed study of the geology of coalmining would be out of place here; but some of the results of earth movements must be mentioned. First, coal seams, which were horizontal when they were laid down, are almost invariably inclined. In some localities, (eg parts of Scotland, Lancashire, Northumberland, Staffordshire and Wales) this inclination is quite steep,

NATIONAL COAL BOARD AREAS

Fig 1   Map showing the coalfields of Britain and National Coal Board areas

vertical seams not being unknown. Second, in some places the strata undulate. Where the roadways follow the seam, this may make mechanisation of transport difficult. Third, in many places, there are breaks in the strata, called faults. Where a coal-face meets a large fault head on, the coal disappears, lying at either a higher or lower level. The first is called an upthrow and the second a downthrow fault. Where the fault is a small one, it is possible to work through it, but a major fault will stop the face.

In Roman Britain, and possibly earlier,[3] coal was dug from surface outcrops. Outcropping on a small scale was practised in medieval Britain, too, but in places where the surface deposits became exhausted coal was then extracted in one of two ways. Where a seam outcropped on to a hillside, the miners tunnelled into it. These early drift mines were very shallow because of ventilation and drainage problems, but by the fourteenth century these problems were partly overcome by the driving of adits, and it was therefore possible to penetrate more deeply into the hillsides. Where the ground was flat, the other early method was to sink bell pits. A bell pit is essentially like a well sunk to a shallow seam of coal. The miner dug the coal from the bottom of the pit, and hacked into the sides for as far as was safe. He then abandoned the pit and sank another close by. Viewed in section, such a pit is usually shaped like a bell.

Plate 1   Bell pit exposed on an opencast site at Carlcotes, Yorkshire. *(E. A. Dyson)*

In drift mining, the practice developed of driving short side headings into the coal at intervals along the main heading. The length of these side headings (commonly called stalls) was restricted by lack of ventilation, but in the course of time miners found that coupling two stalls together by cutting a passageway between them created a flow of air. This process could be repeated until the mine consisted of stalls from which the coal was won, separated by rectangular pillars of coal left to support the roof. This system of working is called stall and pillar, and it has a number of variants the most important one being the bord and pillar of Durham and Northumberland. Similarly, where coal was won from shafts, bell pits gave away to stall and pillar once the very shallow coal was exhausted.[4]

The equipment used in these early mines was simple and cheap: hand windlasses, boxes and baskets to hold the coal, picks, wooden shovels, maybe wooden mauls (mallets) and wedges to break down the coal, buckets for clearing water out of the workings, and little else.

As to location, in the later medieval period there were coalmines in all the coalfields being worked today, except for Kent.[5] These early workings followed the line of the outcrops of the coal seams. Mining was on a small scale. Many peasant farmers got a little for their own domestic use. In other cases, a small group of men produced coal for the market. But it was a restricted market. Since few houses had chimneys before the late sixteenth century,[6] wood was generally preferred for domestic fires, and there were not many industrial uses for coal. Furthermore because of the cost of transporting coal by packhorse, the market for it was a local one. William Gray, referring in 1649 to the North Eastern coal trade, said:

> Which trade of coale began not past fourescore years since. Coales in former times was onlly used by smiths, and for burning of lime; woods in the south parts of England decaying, and the city of London, and other cities and townes growing populous, made the trade for coale increase yearely, and many great ships of burthen built, so that there was more coales vented in one yeare, then was in seven yeares, forty yeares by-past; this great trade, hath made this part to flourish in all trades.[7]

There can be no doubt that the demand for coal expanded rapidly from about the middle of the sixteenth century stimulated by the growing scarcity of timber which caused its price to rise faster than that of other commodities. Increasingly, coal took the place of wood on domestic hearths; and various trades found ways of substituting coal for wood or charcoal. Evelyn, in 1661, wanted to banish all brewers, dyers, soap and salt boilers as well as lime burners, down below Greenwich so as to rescue London from that 'hellish and dismall cloude of sea coale', and this gives us some idea of the widening range of demand for coal. Nef estimates that there was a fourteen-fold increase in the average annual output of coal between 1551-60 and 1681-90: 210,000 tons to 2,982,000.[8]

Plates 2 and 3   Primitive small-scale working during the 1926 lockout, between Langsett and Stockbridge, Yorkshire; *(above)* a drift mine, *(below)* a shallow pit. *(E. A. Dyson)*

The North Eastern coalfield of Northumberland and Durham felt the pull of the London market because coastal shipping provided cheap transport. Already, by 1649, this coalfield had to draw its corn from far afield, from the southern as well as the northern counties of England, from Scotland, and from Denmark.[9] Other coalfields near the coast, Cumberland, parts of Scotland, North and South Wales also despatched coal by sea, the relatively small Cumberland field in particular benefitting from her proximity to Ireland. But none of them rivalled the North East for size nor for the degree of specialisation in coal as against agriculture and manufactures.

In the case of the inland coalfields, because of the very heavy cost of transport by road, only those mines which were near a growing town or navigable river (the Wye, Severn, Mersey, or Trent), expanded to any great extent. These coalfields remained areas of mixed economic activity, with coalmining remaining an adjunct to agriculture. Nevertheless, many of the innovations in mining technology were developed in the Midlands. Wollaton, in the early seventeenth century, was the most technically advanced colliery in Britain. Huntingdon Beaumont, who managed it, was responsible for introducing to the North East:

> . . . many rare engines, not then known in these parts; as the art to boore with, iron rods to try the deepnesse and thicknesse of the coale; rare engines to draw water out of the pits; waggons with one horse to carry down coales, from the pits, to the stathes, to the river, etc.

Gray's description of North Eastern coalmines in the mid-seventeenth century indicates that they were at the same level of development as many in the Midlands. Coal, loaded into corves (hazel baskets), was drawn along the roadways underground by boys. In the pit bottom, the corf was 'hooked-on' to the hempen winding rope to be drawn out of the pit by horses. However, on the whole these mines were deeper and larger than those in the Midlands. Gray implied that some coal was being worked at a depth of 40 fathoms (240ft) and from archaeological evidence it can be concluded that few, if any, Midland examples rivalled this. Again, he speaks of many thousands of people being engaged in producing and transporting coal in the district, while one 'coal merchant' alone employed over 400 in his coal works.[10] Even allowing for exaggeration, the comparatively large scale of the operation is clear. Further, while the pitmen of Tyneside were mainly full-time underground workers, in other districts most were engaged in mining for only part of the year, although colliers in the North East sometimes worked on the colliery farm as late as 1750.[11]

So far, expansion in output had not been accompanied by any very startling improvements in techniques or equipment. Exploitation of the coal measures had become more extensive, but not more intensive. A more intensive exploitation could only follow from improved systems of drainage and ventilation, and these did not emerge until the eighteenth century.

The steep increase in the demand for coal in the seventeenth century was followed in the first half of the eighteenth by a steady growth in the exporting districts and stagnation in inland coalfields; and then by a further steep increase from about the middle of the eighteenth century. As against the fourteen-fold increase between the middle of the sixteenth century and the end of the seventeenth, Nef postulates a three-fold increase in the eighteenth: from 2,982,000 tons a year in 1681-90 to 10,295,000 tons a year in 1781-90. Most of the increase took place in the second half of the century. Thus, coal imported into London, mainly from the North East, averaged 411,000 tons a year in the decade 1700-9; 599,000 in 1741-50 and 1,081,000 in 1791-1800.[12] By 1800, the inland coalfields were expanding output rapidly in response to the increased demand resulting from the growth in the population and the development of manufactures. So far, they supplied only a trifling share of the London demand; but the growing network of canals provided them with a wide market area. The coming of the railways would enable them to compete in London too.

From the Industrial Revolution until the early 1920s the coalmining industry expanded almost continuously, until by 1923 there were 1,160,000 miners who produced a total of 276 million tons of coal, just short of the record output of 287 million tons of 1913. From this point the numbers employed and the output fell steadily and by 1945 there were 709,000 miners who produced 183 million tons of coal.[13] Since then, output and demand have fallen still further; but there are now signs of a resurgence.

1
The Organisation
of the Industry

THE RIGHT TO WORK COAL

Most of the coal used in Britain in the Middle Ages was dug from surface deposits in much the same way as turf and peat. The manorial tenant's right to take it for his own use was governed by the custom of the manor on which he lived. Generally, so long as coal was available on, or no more than a few feet below, the surface his right to take it went unquestioned. However, as coal became a saleable commodity and it was necessary to dig drifts or pits to reach it, disputes occured between lord and tenant as to their respective rights.[1]

As in other respects, the freeholder was generally in a much stronger position than the copyholder, but there were cases (for example the Stanleys of Hawarden Castle, Flintshire) where the lord of the manor claimed the right to work coal under the land of his freehold tenants, sometimes not even paying compensation for the surface damage caused by it. More often, freeholders were induced to surrender their right to work minerals in return for compensation which was frequently inadequate. In other cases again, lords refused to grant wayleaves, so that their freeholders had no way of getting their coal to market.[2]

Generally, copyholders (tenants holding land by copy of the manorial court roll) lost their right to take coal once the surface deposits were worked out, and lords rarely had difficulty in establishing that they were entitled to work minerals under copyhold land. However, the lord was usually obliged to make good or pay compensation for surface damage; and in some cases copyholders were given an allowance of free or cheap coal in lieu of 'firebote': the medieval right to take coal for their own use. The right of the copyholders of Eckington to 'work and sell the coal and mines' subject to the proviso 'that the sale thereof does not hinder any sale by the lord' was altogether exceptional by the seventeenth century. It continued until the nationalisation of freehold coal rights in 1942 when owners of former copyhold land there were, in effect, compensated as though they were freeholders. A similar custom existed on a Durham manor where copyholders had the right to work the upper coal seam.[3]

Some attempts were made both in England and in Scotland to establish the Crown's ownership of base metals and other minerals as well as gold and silver, but they were not sustained. The right of the subject to such minerals

16

lying under his freehold land was established by the courts, especially in *Regina* v *the Earl of Northumberland,* 1566. In Scotland, an Act of 1424 tacitly recognised that the Crown's mineral rights applied only to gold and silver; but a further Act of 1592 purported to bring 'copper, lead, tin, and other whatsoever metals and minerals' into royal ownership. Landowners were to be required to pay to the Crown 'the just tenth part of all and whole the said gold silver copper lead tin and other minerals which shall be found and gotten yearly within their said lands and heritages', but the act was ineffective. In Scotland, as in England, the ownership of minerals other than gold and silver was vested in the freehold owners of the soil.[4]

Nevertheless, there were some local customs which ran counter to this general rule. The principal case affecting coalmining was the Forest of Dean. Here, a man born within the Hundred of St Briavels who had been a miner for a year and a day, on finding coal or iron ore could apply to an officer of the Crown called the gaveller for a 'gale': an area of land under which he could work the mineral. Originally, the extent of a gale was determined by the distance to which the man could cast soil from his mine, and this was appropriate to bell pit working: but the size of the gale was increased several times as mining technology developed. While the free miner had possession of, and the right to work, the coal within the gale, he was not the absolute owner of it. He was required to make royalty payments to the Crown and also, if the coal lay under the freehold of a private individual, to him as well. The Crown's share was based on one-fifth of the gross profits; a commutation of an older practice where the gaveller could insist upon adding a fifth man (the king's man) to a partnership of four free miners.[5] Attempts to take away the rights of the free miners in the seventeenth century were strongly contested, and indeed these rights have been largely preserved down to the present day. However, with the development of large-scale mining requiring much capital, it became increasingly common in the nineteenth century for free miners to assign their gales to capitalist firms.

Another area with its own customs was the Honour of Peverel, which covered much of Derbyshire and Nottinghamshire. Here, the Crown claimed ownership of coal lying under manorial waste unless the lord could prove a grant under the great seal. Elsewhere, lords of the manor generally claimed ownership of coal under the waste. There were a few exceptional cases like Bolsover (Derbyshire) where freemen or sokemen had the right to dig coal from the waste. Again, in the charter which he gave to Swansea in 1305, William de Brows authorised the burgesses to take 'carbonem terreum' (earth coal) in Ballywasta for their own consumption, but they were not to sell to strangers; and the burgesses of Holt (Denbighshire) were given a similar concession in the fifteenth century.[6]

Finally, mention should be made of coal gathered on the sea shore. Where seams outcrop on to cliffs washed by the sea, coal is eroded; pieces carried out to sea are later brought in by the tide. In Northumberland the gathering of coal from the shore by local people has been practised since the Middle Ages. It used to be thought that the term 'sea coal', used to distinguish coal from charcole in early times, was derived from this practice. However, it is almost certain that sea-coal was so called because it was transported by sea.

So far, we have concentrated on the small producer getting coal largely for his own consumption. A distinction may certainly be made between the Northumbrian villager picking coal from the beach, or the copyholder digging enough to boil his pot, on the one hand; and producers like the free miners of the Forest of Dean, many of whom undoubtedly produced for the market, on the other. There can be little doubt that by the sixteenth century, there were, in every coalfield, small men, whether freeholders, copyholders or leaseholders, producing for a local market and not merely for their own use.

An early example, dating from 1316, of a company of colliers leasing a mine provides for Adam, son of Nicholas, with eight other men of Cossall (Nottinghamshire) to work a pre-existing coal mine paying the owner, Richard de Willoughby, 12d a week for each 'pickaxe'. They were excused payment when unable to work through floods or gas.[7] Clearly, these men were producing for the market, but such an enterprise calls for little organisation. The owner provided the fixed capital, shaft, windlass and sough (drainage channel); but the colliers would need to market their coal (either taking it to Nottingham themselves or selling to a middleman), and to provide the circulating capital. Some of the revenue left over after paying the rent would have to be put on one side to meet materials costs, for example. The fact that Adam was named probably indicates that he was the recognised leader of the partnership who paid the bills, kept the accounts, and organised the work.

The rent of 12d a week for each man seems high, being equivalent to approximately 4d a ton, probably a third of the selling price. Willoughby doubtless considered that this would provide him with a higher and more certain income than working the coal by direct labour as many estates did.

In the Middle Ages, coal was undoubtedly worked in many cases by unfree labour. We find an echo of this as late as 1496-7 when Leonard Forester, surveyor of labourers and workmen for the Bishopric of Durham, had power to imprison the disobedient.[8] Getting outcrop coal was very little different from labouring in an arable field, so where villeins were required to provide labour services as rent for their holdings, they might well be put to work sometimes in the coalfield. It was suggested to the Midland Mining Commission (1843) that the buildas system, where miners were required to do some work in return for an allowance of beer, was an echo of medieval boon work. This is not to suggest that the Durham miners of the late fifteenth century were villeins, however. While there was, it appears, some element of bondage in their situation, they were working for wages with coalmining as their regular occupation.

The Great Northern coalfield of Northumberland and Durham was already feeling the pull of the London market. In contrast with the Cossall lease of 1316 we may take a Durham example of 1356. Here, Bishop Hatfield leased out his Bishopric's five mines at Whickham for 12 years at a rent of 500 marks ($£333$) a year.[9] Here again, an element of bondage is indicated by a provision that the bishop will not take any of the lessees' men away without the lessees consent, which indicates that he had the power to do so if he wished.

This lease contains a clause limiting the permitted output to 1 keel (about 20 tons) a day from each mine. Having a fixed annual rent, but limiting the output (by specifying either a maximum output or a maximum number of men to be employed) was common in medieval and early modern leases. For example, an early lease (dated 1322) permits the lessee, Walter de Coldebrok, to employ one man only to dig coal in 'Le Brocholes' (brookholes) belonging to Wenlock Priory for a rent of 6s for one year. Similarly, a deed of 1326 relating to the same district permits Adam Peyeson a tenant of Hugh, Lord of Scheynton, to employ four men to dig 'Secoles' on the land he holds. There were exceptions, however, as in the case of two collieries at Elswick owned by Tynemouth Priory which were let in 1330 for 100s and 80s a year respectively without any limitation on the output.[10]

By the seventeenth century, it had become the general practice to specify an amount of rent per unit of output: so much per load, or per ton for example. Unfortunately, there was a bewildering variety of units and reducing them to imperial tons involves a degree of guess-work. Seventeenth century leases usually provided that lessees had the right to take timber from the estate, and the duty to keep the sough (drainage channel) in repair. Provisions as to the making of roads to carry the coal away, wayleaves, and so on were also common.[11]

On the whole, early mining leases were for short terms. The company of colliers at Cossall in 1316 were virtually tenants-at-will with hardly any security of tenure. Religious foundations (which owned most of the Great Northern coalfield before the dissolution of the monasteries) also tended to give short leases; monastic houses in particular usually charged high rents for them, so their lessees had no incentive to invest in capital works. Even in the late eighteenth and early nineteenth century, 21 year leases were common. Marx suggested that short leases were responsible for the inadequacy of the houses built for their workers by mining entrepreneurs. As always, there were exceptions. One lease of 1354, for example, was for 30 years; again, in 1687 the Bishop of Durham leased mines to Thomas Langley of York '. . . to have and to hold for the longest of three lives'; a Nottinghamshire lease to the monks of Beauvale Priory in 1457 was for 99 years; while in 1654 the governors of St Bee's Grammar School, Cumberland granted a lease of 867 years duration.[12]

The period 1540 to 1640 when prices rose markedly, also saw an increase in royalty payments which were, on the whole, high by comparison with the early twentieth century. In the late seventeenth century, they averaged about 5*d* a ton, although there were some as high as a shilling: that is a quarter or more of the selling price. From the middle of the eighteenth century, the general level of royalties tended to fall. According to Adam Smith, writing in 1776:

> The rent of an estate above ground commonly amounts to what is supposed to be a third of the gross produce; and it is generally a rent certain and independent of the occasional variations in the crop. In coal-mines a fifth of the gross produce is a very great rent, and it is seldom a rent certain, but depends upon the occasional variations in the produce.[13]

In the 1830s, royalties in Northumberland and Durham averaged about one-fifth of the selling price. In 1918, for Great Britain they averaged about one-thirty-fifth, although this masks wide regional variations. Thus, for the second quarter of 1918, royalty payments averaged between 4*d* and 4½*d* in the Midland counties compared with just under 10*d* in Scotland and Northumberland and almost 1*s* 3*d* in Cumberland. The drop in the proportion of the total selling price taken by royalty owners between the mid-nineteenth century and 1918 can be simply explained: the selling price of coal rose very considerably while the royalty per ton remained virtually unchanged.[14] The pressure of public opinion, which was antagonistic to royalty owners was no doubt partly responsible for this.

Of course, a royalty payment is not purely rent. First, the medieval lord expected his share of the produce from the land. In Somerset this concept never quite died out. Early royalties there were called 'free shares' amounting to one-eighth or one-tenth of the value of the coal, and there is at least one twentieth-century example of a 'free share' equal to one-twentieth of the value, which was distinguished in the accounts from the royalty rent proper, paid in addition. Second, in modern times it was recognised that winning coal removed an asset from the land which could not be put back. Again, in many early leases, the lessee worked an existing mine in which his landlord had invested capital for shaft sinking, drainage and possibly machinery. The royalty therefore was partly a return on this investment. By contrast, in later leases the landlord invested little or nothing. For example, an elaborate lease of 1738 from Squire Edge and others to the Barber Walker partnership at Strelley, Nottinghamshire, envisages that Barber Walker will sink pits, erect engines, make drains and ways, and so on. Edge has no responsibility for providing anything. This lease is, for its period, unusually long-term: 99 years, and the rent payable of 1*s* 3*d* per stack load, equal to about 6*d* a ton, about average. This lease does not provide for the payment of a 'rent certain', that is minimum rent to be paid irrespective of output, but most eighteenth century (and later) leases did so.[15]

This consideration of mining leases should not be allowed to obscure the extent to which landowners worked their own coal, either by direct or sub-contract labour.

THE EMERGENCE OF THE ENTREPRENEUR

Before the Reformation, a suprisingly high proportion of land in coalmining districts was owned by religious institutions whose policy was, on the whole, conservative. The price inflation of the Tudor period coincided with the transfer of much churchland to the Crown and private individuals. Now, with the demand for coal rising rapidly, there was every incentive to expand coal output, and the change in ownership of so much coal-bearing land facilitated expansion.[16] One of the most important transfers was of minerals in Gateshead and Wickham from the Bishop of Durham first to the Crown, and subsequently to the merchants of Newcastle, in the early 1580s. This lease by the partnership of Newcastle merchants, in which there were some two dozen important partners, was known as the 'Grand Lease'. In 1590, the Lord Mayor of London complained to Lord Treasurer Burghley that coal prices had risen from 4s to 9s a chaldron since the negotiation of the 'Grand Lease' which had given the lessees a virtual monopoly of the coal trade on the Tyne. In 1600, the Newcastle traders' monopoly was formally incorporated by royal charter, as the Company of Hostmen. The Company had thirty-one members in 1622, operating mainly on their own capital, but many other local traders had a small stake in their fortunes. Whereas in other coalfields most of the capital invested in coalmining continued to come from the land, on Tyneside much of it now came from trade. Besides the 'Grand Lease', there were many other partnerships of traders working collieries on both banks of the Tyne, and on the Wear, in the first half of the seventeenth century. In many cases, the partnerships had both landowning and merchant members. According to Nef, there were few large collieries in either Nothumberland or Durham being worked by 'a single capitalist' by the Civil War; although there were, of course, still some being worked by individual landowners on their own freehold.[16]

Because the members of a seventeenth-century mining partnership in Northumberland and Durham were primarily traders, they received their shares of the product in coal. This was tipped on to separate heaps on the pit bank, according to the respective shares of the partners who were then generally responsible for arranging their own sales.

The principal hostmen, consisting of twenty or so Newcastle merchants, controlled both the production and sale of coal. The lesser members of mining partnerships (whose shares, being often divided between surviving relatives, become more fragmented with each generation) were called upon to pay their share of the expenses but exercised no control over the management of the collieries or the trade. The same small group of trading families had controlling interests in many theoretically separate colliery undertakings. They could thus decide that one mine should expand its production, while another contracted or even stood idle, however detrimental this might be to the interests of the inactive partners. Further, since their charter of 1600 gave hostmen a monopoly of coal trade on the Tyne, they could ensure that no else could profit from investing in mines in the hinterland of the river. However, with the slump in the coal trade which coincided with the Civil War, some hostmen began to trade in the coals of non-members and finding this more profitable than producing coal themselves, withdrew from the ownership of collieries. It is not without significance that Gray's *Chorographia,* which gave warning of 'the uncertainty of mines, a great charge, the profit uncertain' was written in 1649. Gray did not rate very highly the chance of profit for: 'Some Londoners [who had] of late disbursed their monies for the reversion of a lease of colliery, about thirty yeares to come of the lease'. In his view, when they came to 'crack their nuts' they would 'find nothing but the shells'.[17]

Despite the importance of merchant capital in Durham and Northumberland, the mining industry continued to be dominated by landlords until the early nineteenth century. This was particularly true of Scotland, but in England families like the Lowthers of Cumberland (who, for long, enjoyed the lion's share of the trade with Ireland), the Willoughbys of Nottinghamshire, the Orrells of Lancashire, the Fitzwilliams of South Yorkshire and the Dudleys of Staffordshire to name but a few of the more important, played leading roles in their respective districts for one generation after another, and this is equally true of the Durhams and Londonderrys of Northumbria. Working the minerals under their own land, they had no rent to pay to anyone else; and, in an era of short leases, only they could be certain of continued enjoyment of capital works like soughs, without which many pits could not work.

Sometimes, members of landed families themselves leased coalmines from others, or entered into partnerships to work others' coal. The Beaumonts of Coleorton in Leicestershire, for example, leased mines at Measham, and others in Warwickshire. The most adventurous member of the family, Huntingdon Beaumont, attempted in partnership with Sir Percival Willoughby to obtain a monopoly of coal supplies in the Trent Valley in the first two decades of the seventeenth century, bankrupting himself in the process. He tried to retrieve his fortunes by an adventure in the Great Northern coalfield in association with a group of London merchants, but after losing his money — £20,000 according to the exaggerated estimate of Gray — he 'rode home upon his light horse'.[18] However, he made his mark on the district by introducing various 'rare engines' and giving his name to a coal seam, still known as the Beaumont seam. Although a member of a landowning family, Huntingdon Beaumont had neither land nor any great store of capital. He performed an entrepreneurial role. Another country gentleman who had become a mining entrepreneur in a district other than his own was Sir Humphrey Mackworth, son of Thomas

Mackworth of Betton Grange, Shropshire. In 1690, he moved to Neath where he married into the Evans family which dominated the coalmining industry there. He invested heavily in the mines, but his chief contribution to the development of the enterprise was his energetic and skilful management. He laid railways from his pits to the coast, and fitted sails on the wagons for use when there was a favourable wind. Sir Humphrey is also almost unique in having recruited seventeen condemned prisoners to work underground; they were bound for five years and their sentences were revoked subject to the honouring of the bond.[19]

Mining entrepreneurs became increasingly important during the eighteenth century. Taking the Erewash Valley, which covers the Nottinghamshire and southern half of the North Derbyshire coalfield, there were in 1739 upwards of twenty collieries, most of them owned by landlords and many of them at that time standing idle. Of the mines at work, the Willoughby's Wollaton Colliery, which was close to Nottingham and the Trent, was probably still the largest. Other collieries being worked by their landlords, one at Smalley by Richardson, another at Denby by Drury-Lowe, and several at Ilkeston by the Duke of Rutland, were comparatively small, being further away from Nottingham and the Trent and so having much higher transport costs except in the restricted local market around the pits. In competition with them was the partnership which came to be known as Barber Walker, working mines at Smalley, Bilborough, and Kimberley. The original members of this partnership were yeomen who had worked coal on a small scale on their own freeholds before leasing mines from others: for example Drury-Lowe at Denby, Edge at Bilborough and Lord Stamford at Kimberley. Their earlier leases were of mines already sunk although they certainly improved some of them by driving soughs. These leases were for short terms, eg one life in the case of Denby. In 1738, they took a ninety-nine year lease of minerals in Strelley. To exploit this coal, they needed a considerable capital, much of which came from the profits of their earlier mines. The care taken in drawing up the lease so as to protect the partnership's interests at the end of the ninety-nine years exemplifies the emergence of a firm. This is in contrast with the typical landed proprietor's attitude.[20] For him his coalmines were part of his estate, yielding current income in the same way as his farming activities. Lord Donington, owner of the Moira collieries in South Derbyshire may be taken as a late example. He purchased the mines from the estate of his brother-in-law, Lord Hastings, in 1870, for about £250,000. He had landed estates in Derbyshire and Scotland, and maintained three large houses: Donington Hall, Derbyshire, Hungerford Castle, Bath, and a mansion at Westminster. His farm rentals and colliery profits fell greatly after 1873, a period which has been styled the Great Depression. In order to maintain his extravagant way of life, he economised on materials, development and maintenance, so as to maximise his current revenue. In consequence Moira, from being one of the most efficient concerns in the East Midlands, became increasingly dilapidated, until after Donington's death in 1895 when a limited company bought it.

The important coalmining district of south-west Lancashire received a substantial injection of merchant capital at an earlier date than most inland coalfields. According to Langton, in the 1770s most collieries were run and financed by the local gentry and mining partnerships of long standing, whereas 'the rapid growth of the 1780s and 1790s was accompanied by a massive influx of capital from Liverpool and Bradford.'[21]

In the nineteenth century, partnerships and, in the second half of the century, joint-stock companies, proliferated and yet there were still many sole owners. In 1867, there were for example 281 colliery undertakings in the Manchester Inspection District of Lancashire, and of these, nineteen belonged to the Duke of Bridgewater's trustees, one (consisting of three collieries) belonged to the Earl of Bradford, eight belonged to James Deardon, three to the executors of Thomas Duckworth, five to Thomas Fletcher, eleven to the executors of James Hargreaves and three to Robert Lees. There were, besides, many sole proprietors of one or two collieries. On the other hand, only six limited companies are listed.

Taking another district at random, the Earl of Durham, Earl Vane and Sir C. F. Maclean each had several collieries in South Durham, and a number of other sole proprietors are listed.[22]

Many sole proprietors employed only a handful of men. One of the surprising features of recent mining history is, indeed, the extent to which the small enterprise has been able to co-exist with the large. One reason for this is that the exploitation of shallow coal reserves in past centuries has been patchy, and there are many odd pockets of coal near the outcrop which can be easily worked using primitive equipment. The owner or lessee of such a small mine needs a local market for his output, and he can usually ensure this by charging slightly less than his competitors. Throughout the nineteenth century, such small mines were considerably more numerous than large ones, but they supplied an insignificant part of the total output. Even so late as 1945, there were still 521 coalmines in Britain employing less than 30 wage-earners. However, while they comprised one-third of the total number of mines, they supplied only one per cent of the total output.[23]

THE MANAGEMENT FUNCTION

The management of a small mine supplying a local market, whether in the fifteenth century or the twentieth, involves no great problem, and we need not let it detain us here. However, where the small mine owner or lessee was unable to market his own coal, he might well find himself dependent on, or subordinate to, a middle man. There is no doubt that some of the companies of colliers working mines in the late Middle Ages found themselves in this position. Similarly, the small producers on Tyneside were unable to ship their coals after the Company of Hostmen was formed.

Early coal workings were regarded as an estate activity, a kind of specialised agriculture. In wooded districts, coal could be worked from shallow drifts and bell pits without disturbing the trees overmuch. It is therefore not surprising to find the monks of Durham Priory appointing a 'Forester' to supervise their colliery at Rainton in 1367. Similarly, the coal works of the Bishopric of Durham were part of the province of the master forester until a surveyor (or chief viewer) of mines was appointed in 1384. Another early title for a colliery official was 'banksman'. One Thomas Buk was appointed to this position by the Bishop of Durham about 1438, before being promoted to surveyor at a yearly salary of 40s. It appears that the surveyor was originally the person having overall charge of the mineral estate, carrying out periodic inspections while the viewer exercised more direct supervision and the banksman kept a check on the amount produced and sold, and had authority over the underground workmen, a function subsequently filled by the overman. The term 'banksman' meaning the person in direct charge of a mine, was little used in the Midlands, although there is one late eighteenth century example. John Morton, 'banksman' at the Bramley Moor Colliery, owned by the Sitwells, managed the colliery and rendered accounts of the usual master and steward single-entry kind.[24]

Until late in the nineteenth century, colliery managers were known as viewers in Northumberland and Durham, and the title can perhaps be better understood if we think of the viewer being originally an organiser and supervisor of a chain of bell pits or drifts in wooded country. Where the pits were leased out, he exercised supervision ('viewed') on behalf of the landlord, making sure that the mines were being worked in a 'workmanlike manner' and that the proper amount of rent was being paid.

Once the very shallow coal in an area was worked out then a new skill was called for: the skill of finding coal at a time when no one knew anything very much about geology. The 'searching out of coals' was a mystery rather than an art until Huntingdon Beaumont pioneered the use of boring rods early in the seventeenth century. Perhaps Beaumont may be taken as representative of a colliery manager of a new type. He was, first of all, an engineer. Besides his drill rods 'to try the deepnesse and thicknesse of the coal', he also introduced into the North East improved pumping engines and railways. Wooden railways, subsequently known as 'Newcastle roads' were modelled on the one Beaumont laid down from Strelley to Radford, Nottinghamshire in 1604.

Then, Beaumont designed the new pits, appointed and supervised the subordinate officials, negotiated leases, sold the coal, bought the materials, and so on. One surprising feature of his correspondence with Sir Percival Willoughby is the considerable technical detail they discussed in this way, with Sir Percival showing knowledge as well as interest. For example, Sir Percival writes:

For the fludds I doe not greatelye feare them, for the passages to the soughe be daylye looked unto and in tyme and measure kepte, wiche you knowe ar everye hower subiect to slipps and falls, and especiall in fowle weather if they be not sure before, there is fall inoughe. Yet I thincke it not amisse as you write, to sett downe a pump in the pitt that leadeth to the horse water, but I doubt there will not be roome inoughe for anye quantite of water to lye there, neither will one pumpe take all the water that is cast up from the myne. If the pitt be bigg inoughe, to sett downe two pumps and that one cogg whele maye turne bothe the ragg wheeles. . . .[25]

The detailed supervision of labour was undertaken by subordinates. 'Stovers of the field' had to give a just account of what they had 'gett and sould' every noon and night according to rules codified in 1600. The stover (a term used only in the Erewash Valley) was the senior partner in a sub-contract arrangement. He managed the pit, keeping a check on the work of the underground men and making sure that the owner received the right amount of money for the coal produced and sold. Subordinate to him was an 'underman' (later called a 'butty') supervising the colliers underground. At Coleorton Sir Francis Willoughby was paying wages on the sub-contract system as early as 1570.[26]

Perhaps it would be convenient here to deal with the development of management in the North East, because this foreshadowed what was to happen elsewhere. By the seventeenth century, the most senior manager of a large undertaking was usually called the viewer, although an alternative title was 'head Under-Over Man'. Under him was an overman or banksman, or both. Where there were several pits, there would usually be only one viewer for the whole undertaking, but an overman for each pit. Sometimes, these overmen worked the pit on contract, but more often they were salaried officials. Some small undertakings managed without a viewer, the overman then being responsible directly to the owner.

A treatise on coalmining published in 1708 specifies the duties of the viewer. He has charge of three or four working pits for which he expects a salary of at least 15s or 20s a week: more if he has more pits to supervise. (A hewer on day wage at this time was paid 12d or 14d a day, so the viewer was paid about four times as much as the hewer). The viewer plans the workings, putting on lines with the use of a compass, decides in which direction the headways shall be driven 'according to the Grain of the Coal, as it lies along the Grain, and not cross the Grain'; decides how wide to make the headways and bords, and how wide the pillars need to be to support the roof. After carrying the workings forward some 200 yards or so at each of the four points of the compass, it is his duty to see that there is a new shaft sunk in time to couple with the old workings. At all times he needs to ensure that there is a good flow of air.

Then, at each working pit is an 'Over-Man' charged with setting the men out and supervising the work underground. On the pit top is an 'Over-Man of the Tree or chief Banck's Man' who keeps an account of the coals drawn out of the pit. If there is any shortfall from the proper amount, this is noted in the banksman's report to the Clerk of the Works who abates the wages of those concerned unless the deficiency is made good on a later day. The banksman also reports on corves (baskets of coal) not properly filled. The hewer responsible for filling a deficient corf is to be reprimanded publicly and if he does not fill another corf to make up the deficiency, is to have sixpence deducted from his pay by yet another official, the 'Steward or Pay-Master'. The steward also dealt with the middlemen who shipped the coal, so it appears that he was the general administrator and cashier.[27]

A diary for 1749-51 kept by Edward Smith, Viewer of Houghton Colliery, Houghton-le-Spring, owned by Mr John Nesham, shows how wide-ranging his duties were. He was at once the manager, mining engineer, mechanical engineer, book-keeper, and paymaster. He also supervised the colliery farm, and sold coal to local customers. He was also consulted on the working of coal by several local colliery owners beside his own employer. He was, it appears, responsible to the owners' agent, Mr Amos Barnes, who visited the colliery only very occasionally. At this date, there were three pits at Houghton Colliery: Hope, Success and an 'engine' pit from which water was pumped.[28]

This pattern of senior management, with a resident viewer acting as general manager and responsible to an agent or steward was usual in the case of estate mines. Edward Smith entered the names of the 'gentleman coal owners' on the River Wear in his diary. These remind us again of the continuing importance of landowning colliery owners in this coalfield:

South Side.
Hen. Lambton Esqr.
John Tempest Esqr.
Mich. Lambton Esqr.
Sir Richd. Hylton Barrt & Mr Nesham

North Side
Sir Ralph Milbanks & Sir Rich. Hylton.
Misstress Allans & daughters
Wm. Peareth Esqr.
Mr Thos. Donnison
Mr Geo. Humble.

One entry in the diary records the overman's task as 'Setting on the pitt', that is, setting the men to work. The hewers, being pieceworkers, would not need very close supervision, but the overman would have to make sure that they did not 'rob' the pillars or do anything else in an unsafe way. His chief concern would be to see that the hewers were not hindered by a lack of empty corves, or bad ventilation, or lack of supports. As mines grew bigger, deputy overmen (deputies for short) had to be engaged to provide these essential services. Thus, at Mr John Nesham's Newbottle Burn Moor Colliery in 1788, there were 10 deputies to 59 hewers, 12 horse drivers, 12 barrowmen and 12 lampmen.[29] However, deputies in Northumberland and Durham were regarded as superior workmen rather than as officials.

The growth in the size of collieries also necessitated the engagement of several officials subordinate to the overman. The 'master shifter' was the official in charge of the repair shift; the 'master wasteman' was in charge of the old men employed in supporting the roof in the waste and keeping airways open, and the 'back overman' was in charge of the second production shift in two-shift working. In 1849, these three officials were paid about 21*s* a week, as compared with the overman's 26*s* to 28*s*. The overman was additionally provided with a substantial house and garden rent free; the others usually had rather smaller houses.

According to Greenwell writing in 1849, the overman

> . . .sets the pit to work each morning, and attends to all the details of arranging the work, and getting the coals each man works to the shaft bottom. It is also his duty to see that each working place is properly ventilated and in a safe state. He also keeps a daily account of the work wrought, and of the whole of the underground expenses and wages, and gives into the colliery office a fortnightly account of the same, the bill containing the amount earned by each man, or set of men if in partnership, and boy during that time . . .
>
> An overman is almost invariably a man who has passed through all gradations of pit work, from the trapper upwards, and who has been raised to his situation on account of his ability and steadiness.

However, Greenwell also notes that at some collieries there was now an underviewer subordinate to the viewer but superior to the overman. Where an underviewer was allocated to a single pit, the overman's status was to that extent diminished.[30] Underviewers in Northumberland and Durham were sometimes promoted overmen, but more often, men who had been trained by viewers, and who themselves would become viewers after a few years' experience. It was already possible to distinguish between the official who was a practical pit man, and the one who had had a formal training.

The evolution of the office of viewer from the estate agent with a particular interest in minerals, to the colliery manager skilled in civil, mechanical and mining engineering and yet retaining much of his land agency function, may be seen as an evolutionary process extending over several centuries. This process was almost complete in the North East by 1850, but in the Midlands it was in a much earlier stage as we shall see.

Even today, land agency tends to pass from father to son. For example, the present agent for the Sitwells' Renishaw estate, Mr Peter Hollingsworth succeeded his father and grandfather. There were countless examples of this process in the sixteenth and seventeenth centuries. It was, again, a natural occupation for an impecunious son of landed gentry like Huntingdon Beaumont.

The skill of mining engineering developed as deeper and more difficult coals came to be worked. This skill was, however, developed from the earlier skill of land management. Thus, J. C. speaks of working according to the grain of the coal by analogy with timber; the sough is no more than a drainage ditch driven underground; the windlass is much the same whether used to wind coal out of a pit or wind water out of a well; and so on. Surveying and calculating the likely product of an area of land has its underground counterpart. Indeed, underground surveys were still being plotted out on the surface by some viewers in mid-nineteenth century.[31] Again, the viewer needed to consider surface damage likely to result from a particular pattern of underground working, and to the hazards of carrying shallow workings under a swamp or pond.

From about 1720, the Newcomen engine came increasingly into use for pumping, and the viewer's mechanical skills then developed. Some viewers became builders of engines themselves. For example, William Brown built his first engine at Throckley in 1756, and many more in the succeeding quarter century.[32] The working parts of these early pumps were made of wood, as far as possible, so there was a natural transition for the craftsman, from millwright or carpenter to enginewright.

The basic education of the early viewer may have been provided by a village-, or local grammar-school; but from the late-eighteenth century many went to private academies, to one of the Scottish Universities or, after its formation in 1838, to the Durham University School of Mines. One of the most eminent of the eighteenth century viewers was John Buddle, 'a person of considerable literary and scientific attainments'[33] who was born in 1743 and who ran a private academy near Tanfield before becoming a full time mining engineer. He was appointed viewer at Wallsend Colliery in 1792 with his son John Buddle junior (who had been educated at his father's academy and who was then 19 years of age) as his assistant. The young Buddle was responsible for some of the most significant improvements in the art of coalmining in the early nineteenth century.

The biographies of the most important colliery viewers in a manuscript volume written by Mathias Dunn show that by 1800 they formed a powerful freemasonry. The unprecedented growth in the demand for coal, when the difficulties in mining it were constantly increasing, placed a premium on their services similar to that noted by Dr Chaloner in relation to civil engineers. In addition to their salaries they enjoyed consultancy fees and perquisites. Many of them were able to purchase shares in undertakings which they managed.[34] Indeed, it is significant that the leading colliery viewer of the mid-nineteenth century, Nicholas Wood, was the first president of both the North of England Institute of Mining and Mechanical Engineers, and the Mining Association of Great Britain (the colliery owners' association) which were founded in 1852 and 1854 respectively. He held both offices until his death in 1865.

Nicholas Wood became a viewer through an apprenticeship to George Stephenson and the two were lifelong friends.[35] Indeed, George articled his son Robert to Wood, a pleasantly reciprocal action appropriate to engineers. Stephenson's career is too well known to require re-statement here, but perhaps we should note that he emerged from the ranks of unlettered colliery enginewrights and was always primarily a mechanical engineer. His brother, Robert Stephenson Snr, also originally an enginewright, became viewer of Pendleton Colliery in Lancashire.

Before considering further the development of professional associations of mining engineers, we should perhaps turn our attention to the situation in other coalfields.

Cumberland needs little special mention. In many respects, it was the North East in microcosm, with deep, gassy pits by the late eighteenth century. The best known viewers of Cumberland, Carlisle Spedding (born about 1696) and his son James, managed the coalmines of the Lowther family at Whitehaven through much of the eighteenth century. Carlisle, the 'fourth son of the principal steward of the Lowther estates', is said to have been sent incognito to the North East coalfield to learn his trade by working as a hewer, something he could hardly have done at Whitehaven. Carlisle is best known for the invention of the steel mill, an early attempt to minimise the risk of explosions although ironically he was killed in an explosion himself in 1755. James is credited with an improved method of ventilation called 'air coursing'. He became a partner in the ironworks at Workington in 1763, while retaining his responsibility for Lowther's collieries.[36]

In Scotland, the supervisor of a colliery was called the 'coalgrieve' or 'grieve'. The grieve was mainly on the pit-top, although responsible for the whole undertaking, and kept the accounts which were of the master and steward variety. He sold the coal, paid wages, collected rents, disciplined the labour force, and so on. Generally, grieves were drawn from the class of estate stewards like their English counterparts, and on small estates the steward acted as coalgrieve himself. Under the grieve was an 'oversman' in charge of the day-to-day operations underground. At a very large colliery he might have 'sub-oversmen' to assist him. There was also a surface official called 'the check' who kept a tally of the amount produced. It will be seen that the grieve was roughly equivalent to the eighteenth-century overman of the Great Northern coalfield, while the oversman's duties were much the same as those of an overman of the mid-nineteenth century.

Scottish colliery proprietors who were not landlords usually had an official superior to the grieve called the 'overseer', roughly equivalent to the viewer. There would be one overseer to a colliery, but one grieve to each pit. The larger estate collieries in the late eighteenth and early nineteenth centuries adopted a similar pattern.

Scottish officials were paid less than their Northumbrian contemporaries. In the early nineteenth century, the grieve might be paid as little as 6s a week (plus perquisites it is true) and the oversman rather less compared with the 15s or 20s or more a week of the Northumbrian viewer, and the 8s of the Northumbrian overman. On the other hand, Scottish colliers who were pieceworkers usually made a small payment to the grieve, which may have introduced an element of bribery. By the end of the eighteenth century, a grieve might be paid about 15s or 17s a week, plus perquisites.

Regarding overseers, some aristocratic coalowners like the Clerk baronets of the Lothians performed this duty themselves. Others emerged from among the grieves, and a few probably came from England. Towards the end of the eighteenth century, overseers of much the same standing as the famous Northumbrian viewers were emerging. One was John Burrel, chamberlain (estate steward) to the Dukes of Hamilton who, in 1767, was paid £50 a year plus perquisites for managing the Duke's coalmines, salt and lime works and woods, and £60 a year, plus a second house and travelling expenses, for other estate work. The overseers employed by non-landed entrepreneurs were often paid no better. For example, Robert Seymour, brought from Whitehaven by Glenbuck Ironworks to manage their collieries and workshops, was paid £105 a year in 1808-10.

Nevertheless, despite low pay, Scottish coalowners were served reasonably well by their officials. One factor here is undoubtedly the superior education available in Scotland; and another, the lack of alternative local opportunities for bright sons of middle-class landed families. Particular reference needs to be made to Alexander Bald of Alloa and his more famous son Robert, probably Scotland's first true mining engineers. Alexander, whose people were griddle makers, became manager of the Erskine family's mines near Alloa about 1774, and he trained his son, and took other pupils too. Robert (1776-1861) practised as a mining and civil engineer from about 1800, 'viewed' many Scottish collieries, wrote the standard work on the Scottish coalmining industry, and acquired an international reputation. Many of Scotland's best-known colliery managers in the first half of the century were trained by him.[37]

As we have noted, collieries in the inland coalfields had, on the whole, a very restricted market area prior to the building of canals in the late eighteenth century, and most of them were therefore small and did not require the same degree of management skill as the larger Northern undertakings. However, the expression 'small mines' conceals degrees of smallness. Some very small mines were managed by working collier proprietors as was one at Staunton Harold, Leicestershire, which was flooded out with tragic results in 1886.[38]

A similar inundation, also resulting from bad management, occured at Molyneux Colliery, Nottinghamshire, in 1869. This mine, employing about thirty men, was leased by Eastwood and Swingler, ironfounders of Derby, who put an unlettered underviewer, Joseph Millership in charge of it, and he was assisted by an equally incompetent deputy, George Churchill.[39] By this time, it was the usual practice of small and medium-sized mines in the East Midlands to have practical pitmen as underviewers, managing day-to-day, but reporting at weekly intervals to a peripatetic viewer. This was not so at Molyneux. Instead, the lessees relied for surveys on John Boot, Lady Carnarvon's agent, who was only concerned with the coal extracted, and his plans were not accurate in relation to the surface.

The Boot family were typical local mining agents. The first member of the family about whom anything definite is known was Eleazer (c1780-1861) who was variously described as a 'mining agent', 'coal agent' and 'colliery manager'. His son John (1801-93) built up the family practice, and John's son J. T. Boot FGS (1835-1914) and son-in-law W. G. Treadwell, assisted him. The Boots acted as agents both to landlords and lessees, and while specialising in minerals they handled all classes of agency work. The Carnarvons of Teversal were their most important regular clients. On the mine management side, Eleazer and John both acted as 'agents' for the owners of small collieries with a wider range of duties than a modern colliery manager would have. John was in charge of Skegby Colliery (with a labour force of about 60) owned by John Dodsley, in the 1840s and his colliery accounts were typical single-entry estate accounts with all money received on the left hand (charge) side, and all money paid out on the right hand (discharge) side. They include some payments for personal expenses of the squire and for the farm as well as for the colliery expenses.[40] John Boot's salary was shown as 21s a week and his father was also on the payroll. Shortly after this John Boot left Skegby and, among other things, was acting as agent for the owner of coal at Strelley, Squire Edge, which was then being worked by Thomas North. Firms like the Boots acquire a great store of information in the form of plans, old reports or accounts of one kind or another, with an admixture of oral tradition and folklore. They were consulted by a number of important colliery promoters on proposed extensions to the Nottinghamshire coalfield, and on at least one occasion advised on a new sinking in Leicestershire. J. T. Boot put down bore-holes in Lincolnshire, to prove the coal measures, in 1875-6.[41]

The Wollaton and Radford collieries of Lord Middleton were sufficiently extensive to warrant the employment of a full-time agent assisted by sub-agents. The agent from 1790 until the 1840s was Charles Chouler and he was responsible to the estate steward, Mr Martin, one of a long line of stewards and agents of that name. When Martin was ill, Chouler deputised for him, taking charge of all the work on the estate including repairs to Wollaton Hall. It is significant that his assistants were still referred to as agents in 1842. The titles 'agent', 'coal agent', 'bailiff', 'ground bailiff' and 'underground bailiff' were used elsewhere in the Erewash Valley at this date, again indicative of the connection with landed estate management. The titles used in the North East: 'viewer', 'underviewer' and 'overman', were hardly ever used in the East Midlands at this date. The most senior official of Barber Walker was, however, called the 'overlooker', reminiscent of Scottish practice.[42]

At most mines in the Midlands, much of the detailed supervision was exercised by butties. The agent was responsible for royalty and wayleave matters, planning, surveying, directing work on shafts, airways and water levels, seeing that machinery was properly installed and supervised, bargaining with the butties, selling the coal and keeping the accounts. But the butty supervised work underground. The agent might insist on the butty working the pit in 'a safe and workmanlike manner', but in many cases left him to his own devices.

In most districts, this system was abandoned from about mid-century, but it lingered on in South Staffordshire and East Worcestershire until well into the twentieth century. There, indeed, butties were sometimes responsible not only for production, but also for ventilation and other 'capital' works. Mr W. N. Atkinson, a mines inspector, reporting to the Royal Commission of 1907 said:

> The Charter Master or 'Butty' system of working the mines is [one] by which the pits are worked by Contractors . . . under an agreement . . . to deliver the coal . . . On the bank or at a wharf, at an agreed price per ton. The agreements are usually terminable by two or four weeks agreement by either party.
> The colliery owner sinks the shafts, erects and maintains the machinery, and pays the winding engineman and boiler stoker. . . .The owner pays more or less for unproductive or dead work underground. . . .
> With the above exceptions, the Charter master [Butty] pays all wages and engages and dismisses the workmen . . . . The Charter master provides horses and fodder, tubs, tools, candles, and explosives; and other stores. Tramrails are provided by the owner, and in some cases, he also provides the timber . . . .

After enumerating the evils of the system, he comments particularly on the pits owned by the Earl of Dudley:

> As matters are at present the management of the pits is far too much in the hands of the Charter Masters. Even when there is a certificated manager and undermanager they sometimes have so many pits they can only give very limited attention to each. Thus, at the Earl of Dudley's Himley Colliery, where there is a manager and one under-manager, there are eleven pits, at seven of which more than thirty persons are employed underground; and at the same owner's Coneygre and Saltwells Collieries, where there is also a manager and one under-manager, they have twelve pits, at ten of which more than thirty persons are employed underground. The pits at these collieries are, in some cases, three or four miles apart, so that it is evident a manager and one under-manager will only be able to devote a very short time to each pit, if all are visited daily; and, in the absence of the manager or under-manager, the Charter Master or his 'doggy' has entire charge of the pits.[43]

Originally, the Staffordshire charter master or butty was in many cases the equivalent of the Erewash stover, and the doggy the equivalent of the Erewash butty (ie the junior sub-contractor supervising directly the underground labour). In time the doggy became no more than a paid servant of the charter master. Sometimes, the term doggy was applied to the man in charge of haulage lads (elsewhere called a 'corporal') but when it became compulsory to have a fireman, examiner or deputy to make inspections for gas, the doggy was often given this to do too.

By contrast with the mines of Lord Dudley, Earl Fitzwilliam's collieries in Yorkshire were managed efficiently. It is possible to criticise the book-keeping (a cumbersome adaption of estate accountancy), but the mines themselves were well run by the standards of the time. The management structure of the estate in the late eighteenth century had, at its apex, the Earl himself, then a hierarchy of land agent, house steward, and clerk (Joshua Biram). Below him each of the four mines was supervised by an overlooker and all four overlookers belonged to the Hague family.[44]

It appears that the 'overlookers' directed operations like the overmen at North Eastern collieries, but with Joshua Biram superintending them from a distance. Important decisions regarding planning, investments, price policy, etc were taken by Fitzwilliam himself on the advice of the land agent or house steward.

In 1797, Michael Hague left Elsecar for a neighbouring colliery in different ownership, taking his colliers with him. The other Hagues also ceased to hold their appointments within a few years, and it seems likely that Fitzwilliam had deliberately decided to change the organisation structure. He brought in one John Deakin to 'inspect and direct' the collieries at an annual salary of £63 when Michael Hague left in 1797; but when Deakin left, his successor was to exercise 'superintendence and keeping the accounts'. By 1807, Joshua Biram had replaced Benjamin Hall as house steward, and he was made responsible directly to Fitzwilliam for the colliery and other industrial functions. The land agent no longer had jurisdiction in such matters. Biram's immediate subordinate was the overlooker Thomas Cooper. He superintended the banksmen (one at each colliery) who were in charge of the surface arrangements including the book-keeping; and he also nominally superintended the underground stewards whose functions were similar to those of Northumbrian overmen. By the time that Benjamin Biram succeeded his father, in 1833, there had been further changes. Benjamin was appointed 'Superintendent and Superintendent of the Collieries'; and at each of the three groups of collieries he had a book-keeper and banksman, and an underground steward at two of the three.

The Birams served their apprenticeship as clerks, but clearly they were able to learn, and were expected to learn, something about practical pit work too. Benjamin Biram was responsible for the 'Management of the entire Establishment', including brickworks, building, farm, smithy, garden, stables, park and house as well as the collieries, but excluding the ironworks. He covered the whole range of management functions: production, sales, purchasing, finance, personnel, and so on. We may particularly note his duty to give 'directions as to the mode of working' the mines, which by 1833 gave employment to over 350 miners and labourers.

Biram's competence as an engineer was called in question by Henry Hartop, who for some years had the management of the Elsecar Ironworks and who made two unsolicited applications to Fitzwilliam for appointment as 'Mining and Mechanical Engineer'. Hartop was probably justified in claiming that Biram's grasp of detailed mechanical engineering matters was inadequate, but there can be little doubt that Biram understood general principles perfectly well, and as a colliery manager his performance was satisfactory. In 1849, Henry Hartop's son, John, again questioned the ability of Biram to direct the work of skilled craftsmen at pits and proposed the erection of central workshops to take over this work. He pointed to the excessive numbers of smiths and fitters at the individual collieries. Fitzwilliam (the fifth earl) seeing the force of his argument, established a central workshop and stores at Elsecar, with Hartop as manager, in 1850. This is still in operation.

Unlike his father, John Hartop was careful not to question Biram's competence as a mining engineer. While it is understandable that trained engineers like the Hartops should have regarded the Birams, who belonged to the older tradititon of specialised land agency, with disdain, the Birams tackled many major mining problems more effectively than their contemporaries. For example, the Fitzwilliam drainage scheme developed in the late eighteenth and early nineteenth centuries kept the workings dry without further capital expenditure until 1920, when the working of deeper seams necessitated modifications. Even so, the modified scheme is still in use today. Benjamin Biram invented the anemometer (used for measuring the velocity of the air current in colliery roadways) a mechanical fan, a safety lamp, and other devices, and Nicholas Wood testified that the Fitzwilliam mines were the only ones, other than those viewed by J. T. Woodhouse, which were well ventilated in Yorkshire in mid-nineteenth century.

Benjamin Biram was consulted by other colliery owners, mining engineers and mines inspectors, and arbitrated in a great many disputes on the working of collieries so there can be no question as to his ability as a mining engineer. One criticism which can be made is in respect of his adherence to the bank work system of coal-getting when East Midlands collieries were going over to true longwall. He may have been influenced in this by the opinions of some Northern colliery viewers (eg Mathias Dunn) with whom he occasionally corresponded.[45]

Three final comments on the Fitzwilliam enterprise may be made. First, while the fourth and fifth earls insisted that the ultimate authority was theirs, the Birams appear to have had their own way with most of the major decisions. Second, it is possible to see the development of the art of colliery management as a special branch of land agency over the period 1795-1850, with Benjamin being much more involved in the practical running of the pits than his father. Third, nearly all the subordinate officials were trained 'on the job', son very often following father.

By contrast with John Boot's 21*s* a week at Skegby in 1847-8, Joshua Biram was paid £150 a year in 1830; and Benjamin Biram, in addition to his salary, had a share of the profit from 1837.

Briefly, we may contrast the situation on the Wollaton estate with the Wentworth estate. The Choulers and the Birams were of the class of land agents; in both cases several generations served their respective masters; in both cases the mines were run as one department of a landed estate; in both cases the officials were trained 'on the job' at their own mines. But Benjamin Biram made the transition to colliery general manager of a recognisably modern type, whereas Chouler did not. And while the Fitzwilliams continued to run collieries, the Middletons withdrew from active participation during the third quarter of the nineteenth century, preferring to become receivers of rents. Despite the basic similarities, there were two major differences between Wentworth and Wollaton. The Earls Fitzwilliam took a direct and continuing interest in the management of their collieries, while the Middletons had a spasmodic interest. The Fitzwilliams saw their function as employers as an extension of their function as paternalistic landlords , and among other things built decent houses for their men, while the Middletons, although humane employers, for long refused to permit the building of cottages for their employees in Wollaton parish because this would increase the poor rate.[46] Again, by the 1840s, the pits at Wollaton and Radford were old, were run on the big butty system (where Chouler had much less to do with the day-to-day running of the pits than Biram) and were facing vigorous competition. A gradual transformation, as at Wentworth, would not suffice here: a revolutionary change was required involving a complete reconstruction of the mines. This did, in fact, follow in 1875 but the reconstruction was undertaken by a joint-stock company.

Like Wollaton, the Moira estate in Leicestershire operated on the butty system, but of a modified type. A butty partnership at Moira had twenty members who contracted to get coal to bank at so much a ton, paid the wages of the day-men and boys and a few other minor expenses, and divided what was left between them.[47] They did not have anything like the authority of butties under the big butty system: they did not engage or dismiss labour; had no control over wage rates or hours of work; did not supply materials (such as timber) and were subject to close supervision by officials.

Like Wentworth, the Moira mines were efficiently run and they came under the direction of the land agent, J. T. Woodhouse, Sen. Under Woodhouse each colliery had its own manager reporting to him regularly, and they were assisted by undermanagers who gave instructions to the butties. J. T. Woodhouse, Sen, besides being the Earl of Moira's land agent, was in private practice as a civil, mining and mechanical engineer from early in the nineteenth century. He trained his brother-in-law, Edward Mammett, and his son, J. T. Woodhouse, Jun, both of whom were renowned consultant engineers.

Unlike Wollaton, the Moira mines were new: the first was sunk in 1804, and five others followed in the next twenty-six years. The second and third earls of Moira were entrepreneurs, who spent money on a number of short-lived and unprofitable ventures including an iron furnace and foundry, a salt works, a brewery, and a spa. The fourth earl, who was a minor when his father died, was, unfortunately, a spendthrift and by the time he died in 1868 he had dissipated a large personal fortune and his collieries were mortgaged up to the hilt. Thereafter, the collieries were in the hands of trustees for a short time before being purchased by another member of the Hastings family, Lord Donington.

The Woodhouses superintended the undertaking until 1878, and despite the depredations of the fourth earl and the foolish economies of Lord Donington the pits continued, on the whole, to be well run until the 1880s. However, F. C. Gillett and S. Smallman, who were general managers from 1878-86 and 1886-93 respectively, were unable to cope with the inevitable tendency to obsolescence in an undertaking which was no longer enterprising. There was little investment in new development underground and new equipment on the surface, and the officials were set in their ways. A new general manager, John Turner, who was appointed in 1893 at what he regarded as a clerk's salary, saw the need for change. He pushed ahead with development work at the expense of current profit, replaced many of the older officials with new men better educated and trained than their predecessors, improved marketing arrangements and tried to persuade Donington to invest in new coal preparation facilities. But Donington refused, considered that the low profit for 1893 (£3,000 in what he thought should have been a boom year) was Turner's fault, and criticised the various changes Turner had made. In fact, the low profit for 1893 resulted partly from a long strike and partly from developing new capacity, which should have been done before. Had Turner neglected to do it, the collieries would slowly have ground to a halt. The major improvements which Turner saw to be necessary had to wait until a limited company took over following Donington's death in 1895.[48]

The influence of J. T. Woodhouse & Co over the development of the coal industry in the Midlands and South Yorkshire needs emphasising. They planned and engineered important new collieries like Cinderhill in Nottinghamshire and Oaks in Yorkshire; they improved systems of coal-getting; and raised engineering standards in all sorts of ways. In the case of Cinderhill Collieries, J. T. Woodhouse, Jun, was the 'consulting engineer' and R. G. Barber (an educated man, a member of a prominent nonconformist merchant family) was the 'resident engineer' reporting to him; and acting under his directions. There was also an underviewer to supervise the day-to-day running of the collieries.[49] Barber was almost certainly one of Woodhouse's pupils. There was a similar arrangement at Oaks, Barnsley, supervised by John Mammatt, a member of Woodhouse & Co. At smaller undertakings for which Woodhouse acted, he would have an underviewer reporting to him.

At any one time, Woodhouse was consulting engineer or viewer for twenty or thirty colliery undertakings spread over all the Midland counties and Yorkshire. True, he had partners (his father; Edward and John Mammatt; J. A. Twigg and others) articled pupils, and various assistants at various times; but he must have found difficulty in giving proper attention to all the concerns for which he was responsible. Nevertheless, his influence was undoubtedly beneficial as one mines inspector acknowledged poetically when he described him as 'like an oasis in the desert in the Midland counties'.

Woodhouse opposed the proposed legal requirement that every colliery should have its own certificated manager. He thought that for 'ordinary sized' as distinct from large mines, it was better to have a peripatetic viewer with overall responsibility because, in visiting many different mines, he acquired wider experience and knowledge than a resident manager. He believed that if the consulting engineer were no longer to have any responsibility, he would not have authority either. His alternative suggestion was that the system of peripatetic viewers should still be permitted, but that more attention should be given to the training of underviewers. Whereas in the North East, underviewers were often well-trained men hoping to become viewers, until the 1880s underviewers (or undermanagers) in the Midlands were almost invariably appointed from among the deputies: they were practical pit men with no great education, and some had only a very rudimentary knowledge of mining science.[50]

The contribution of the North East to the development of colliery management elsewhere is evident throughout the period here discussed. Many viewers trained in Northumbria moved south in search of employment. One of the most famous was John Curr, thought to have been the son of a Durham colliery viewer. He was appointed 'Superintendent of the Coal Works' at the Duke of Norfolk's Sheffield Collieries in 1780, and was responsible for many technological innovations including flanged tram plates, shaft guide rails and flat hemp rope. Unfortunately, he was dismissed in October 1801 because of the poor financial results of the collieries.[51]

Another northern viewer who was engaged to supervise a Midland mining enterprise was Charles Beaumont. His appointment at Earl Dudley's Staffordshire mines lasted for little over a year: March 1797 to April 1798, and it is not clear to what extent the alterations he suggested were implemented. One at least of his technical innovations — the sinking of single shafts divided by a brattice instead of twin shafts — was positively dangerous. He proposed, and for a short time operated, a scheme designed to increase the sale of Dudley's coal, the central feature of which was to supply

26cwt for the price previously charged for 21cwt, the cost to be made good by depressing labour costs. It is not surprising that the colliers rioted.[52]

One of the best known of mid-century northern viewers, George Clemerson Greenwell moved to Somerset in 1853, a year after the foundation of the North of England Institute of Mining and Mechanical Engineers of which he was a founder-member. This body was established to meet at fixed periods and discuss the means for the ventilation of coalmines, for the prevention of accidents, and for general purposes connected with the winning and working of collieries.[53] One of these purposes was to disseminate information and ideas thus raising the general standard of colliery management. Greenwell himself published *A Practical Treatise on Mine Engineering* in 1855 which went through several editions and was widely read by practising colliery officials and students.

In Somerset, he was first appointed manager of Lady Waldegrave's Radstock Collieries at a salary of £500 a year, with a horse and gig, rent-free house, free coal, grazing for a cow and, from 1858, four per cent of any excess of the profits of the collieries over £1,000. He discovered a rich bed of iron ore at Westbury, and became partner in an enterprise set up to exploit it. Lady Waldegrave objected, and shortly after, Greenwell left her employ, subsequently developing a large practice as a consulting engineer, mainly in the Midlands.[54]

Greenwell was followed at Radstock by another North-country viewer, James McMurtrie who, although a Scot, was trained at Newcastle-on-Tyne. He took over the collieries in 1862, and in 1863 was appointed also as agent of the Waldegrave estates. He became a colliery owner himself as chairman of the Great Western Colliery Company and secretary of the Somerset Coal Owners' Association. The output and profits of the Radstock collieries (111,000 tons and £5,936 respectively in 1863) had doubled by 1880. McMurtrie's contribution to this result can hardly be questioned. Radstock Collieries were run as part of the Waldegrave estates until 1925 when another private undertaker, Sir Frank Beauchamp, paid £10,000 for a licence to work them.

Following the Act of 1872, which provided that every mine must be supervised by a manager having a certificate of competency obtained in a state examination, the flow of managers from the North East to other coalfields increased. Perhaps Mark Fryar may be taken as typical of these, although the undertaking he came to manage was no longer typical. After attending Dr Bruce's private academy at Newcastle-upon-Tyne, he trained under a mining engineer named Pole, and at the age of 21 passed the managers' examination. In the following year he was appointed agent and manager of Denby (Drury-Lowe) Colliery, Derbyshire, which had been in the Drury-Lowe family's estate for generations. The Drury-Lowes worked much of their freehold coal themselves, occasionally worked coal under the lands of others, and leased coal and other minerals to other mining concerns.

Fryar therefore had leases and wayleaves to negotiate and enforce; and had to cope with subsidence, drainage and similar incidental problems. One dispute with a neighbouring colliery, Stanley Kilburn, which was not efficiently drained with the result that its water regularly flowed into the Denby (Drury-Lowe) workings, became a *cause célèbre* which was in and out of the high court over a period of nine years from 1885 to 1894. Soon after taking up his appointment, Fryar had to order a new pump costing about £9,000 to replace two earlier ones (an atmospheric engine and another beam engine dating from 1842) and this had to be big enough to cope with the inflow from Stanley Kilburn. He was responsible for drawing up detailed specifications for the pump (though with advice from G. C. Greenwell the consulting engineer) for laying down the engine bed and building the engine house. When the work was completed, there was an official opening performed by William Drury-Lowe and his wife Lady May, which was also attended by Greenwell.

Fryar was, then, the mineral agent, mining engineer and mechanical engineer for the enterprise, but in addition he ordered the materials (giving detailed specifications, and calling for quotations where necessary), canvassed for orders both by writing to old and potential customers and by visiting them; and controlled the labour force. Moreover, besides being responsible for mine rents, he was also responsible for farm, inn and cottage property letting and rents, sales of land, the payment of income tax on the colliery's working, payment of poor rates, highway rates and vicarial titles, and disputes to do with these. On some matters he acted in consultation with Mr Miller, the household steward at the Drury-Lowe home, Locko Park. Finally, when Drury-Lowe stood for Parliament as a Conservative in 1885, Fryar acted as his sub-agent for the Denby district.

Fryar had little assistance on the office side; he handled all the correspondence, but when he was absent on business, a clerk, Alexander Dick, acted for him. On the mining side he had to wait until 1887 for an undermanager (significantly styled 'underviewer' in Mark Fryar's letter book) to assist him. The person appointed was Abraham Bell, previously employed at a neighbouring colliery. He was paid 52s a week, compared with Fryar's £350 a year.[55] Whether this appointment would have been made at that time had the 1887 Coal Mines Act not imposed a legal obligation on colliery owners to employ statutorily qualified undermanagers is not apparent from the evidence available.

Let it be acknowledged that Mark Fryar's position was an untypical one by the late nineteenth century: estate mining enterprises were then very much less important than capitalist partnerships and joint-stock companies where the manager could expect to have specialist help with mechanical engineering, surveying, marketing and so on, and where management was likely to be concentrated exclusively on colliery business. The significance of Fryar's appointment and the way he carried it out is that his training as a viewer fitted him for a post still tied so closely to land agency.

Many men from other districts went to the North East for training in mine management. All the great viewers (and most of the others) took pupils. One such pupil was Charles Morton who in 1850 became one of the first four inspectors appointed under the Coal Mines Act of that year. He was born in 1811. His father, agent to the overseers of the poor in Sheffield, sent him to a school run by a Unitarian minister, the Rev Peter Wright, and then, when he was 13, Charles was apprenticed to J. T. Watter, a Sheffield civil engineer, and subsequently he was articled to the Earl of Durham's colliery agent, Mr Stobart, at Chester-le-Street. After completing his articles, he studied chemistry, geology and mathematics at Edinburgh University before taking up his first appointment as agent for Thorncliffe Ironworks, his responsibility including a colliery near Rotherham owned by members of the firm. In 1837 he became colliery agent for J. & J. Charlesworth, one of the largest firms in Yorkshire, but left them two years later to found a private practice in land and mineral agency at Wakefield. In addition, he became a partner in a colliery concern.[56]

In common with the other mines inspectors appointed in 1850, (J. K. Blackwell, Joseph Dickinson, and Mathias Dunn) Morton saw his role as primarily an educational one. In his first report he said:

> Although it is not within my province to take any direct measures for promoting education among the miners and the subordinate officers, I have availed myself of fitting opportunities to point out to the proprietors and chief directors of collieries the necessity that exists for, and the advantages that would flow from a system of general instruction among their labourers, and of scientific industrial training among their inferior agents; and I have always lent my influence to encourage, and sometimes I have suggested plans for advancing these very desirable objects.[57]

The nineteenth century saw the need for an increasing supply of young, well-trained mining engineers; and for an improvement in the technical knowledge of many existing managers. The requirements of the 1872 and 1887 Acts for all except the very small collieries to have statutorily qualified managers and undermanagers increased the need. As we have noted, viewers in Northumberland and Durham took articled pupils, but similar 'schools' were established elsewhere. For example, John Turner of the Moira collieries was one of many trained by Henry Lewis of Birmingham. There were classes in mining engineering at King's College, London from 1838; and in 1851 the Royal School of Mines was established largely, though not exclusively, to serve the needs of the metalliferous mining industries. The Durham University School of Mines, also inaugurated in 1838, and the North of England Institute of Mechanical and Mining Engineers, formed in 1852, have been mentioned. These bodies did a great deal of educational work leading to the foundation in 1871 of the institution which became King's College, Newcastle-upon-Tyne. Important papers dealing with topics on the frontiers of mining science, many of them reporting on original research, were read to, and debated by, the North of England Institute. This body welcomed members from other districts, and encouraged them to form corresponding institutes in their own areas. In 1889, these district institutes federated to form the Institution of Mining Engineers.[58] Another body, the National Association of Colliery Managers (which has recently merged with the Institution) was formed in 1887 and always concerned itself primarily with educational matters. There is no doubt that many mine managers without formal qualifications improved their understanding of mining engineering principles and how to apply them, through their membership of one of these bodies. For example, Robert Harrison had been a senior official of Barber Walker for 40 years when he said in 1841 that he was 'not aware of any great improvement in ventilating pits, nor does he see how there could be', despite the fact that the pits in his charge had only natural ventilation. Shortly after, Barber Walker installed efficient ventilation furnaces and they were among the first in the district to have mechanical fans. There can be no doubt that the increasing awareness of the need to improve ventilation, and how it could be done, was due in large measure to the educational activities of the North of England Institute and the mines inspectors.

The importance of mechanics institutes, especially for training the subordinate officials, has been stated too often to warrant stressing here.[59] Lord Fitzwilliam, like some other progressive employers, sponsored such an institute for his own employees, about 1840. In the last quarter of the century, evening and Saturday classes in mining subjects organised at university colleges and mining and technical colleges like the one at Wigan, were well attended, correspondence courses were taken by many students (many of them working colliers or deputies who wanted to qualify for the undermanager's certificate), encyclopedic textbooks ran through many editions, and magazines like the *Science and Art of Mining* were in great demand.

One criticism of all this education is that the management of men figured in it hardly at all. To what extent the labour problems of the coal industry have been due to this lack of training in personnel management cannot now be estimated; but it cannot have been inconsiderable.

2

Capital

INTRODUCTION

Small coalmines absorb little capital. Thus, a new pit at Coundon in 1350 cost 5s 6d including 'ropes, scopes and windlass'. A somewhat larger mine, providing employment for five men in the early sixteenth century, cost little more to furnish. A new pit, which took a year or less to work out, cost 2s 6d to 5s to sink; ropes cost 2s 6d each, scopes (ie skips or buckets) 4d each, windlasses are said to have cost only 2d each, and pick sharpening cost 12d a year. The men were paid extra for drawing water, presumably in the scopes.[1] In early mining practice, drainage was the one technical problem likely to entail substantial capital expenditure. In 1486-7, the monks of Finchale spent £9 15s 6d on a horse-driven pump, while in 1552, the Willoughbys of Wollaton spent £1,000 on driving a mile-long sough.[2] At 11s 4d a yard this seems very expensive compared with the 3s 10d a yard Edward Smith paid for a stone drivage at Houghton in 1750, but a drift as long as the one at Wollaton would required ventilation shafts at intervals.[3]

There have been small mines in all districts (except Kent) in all periods, but because coal in Northumberland, Durham and Cumberland was exploited comparatively intensively from about the middle of the sixteenth century, the typical colliery in those counties was larger and deeper than those elsewhere. Coleorton and Wollaton were two of the largest collieries in the Midlands, but even so, we find that in the 1570s five new pits were sunk at Coleorton at around 20s each in a little over two months, so the coal worked was near the surface. Similarly, at Wollaton in the early seventeenth century, several shallow pits were sunk each summer to be worked during the remainder of the year. In Northumbria, shafts were generally deeper, averaging 120 to 180 feet by the end of the seventeenth century, some being as much as 300 to 400 feet. There were also some fairly deep shafts in Pembrokeshire. According to a book published in 1602, they were then being sunk to depths of 70 to 120 feet (costing £20 or more) compared with the 25 feet which had formerly been usual. By contrast, in Leicestershire, as late as 1875, new mines were being sunk to the main coal, a thick seam of good quality, at a depth of only 180 feet. An engineer estimated the cost of sinking two shafts at that date at £1,800 with a further £2,200 for engines, gear, etc. The mine was expected to yield 80,000 tons a year.[4]

Because the sizes of collieries and the complexity of their operations varied so greatly, it is impossible to do more than generalise as to the amount of capital required to finance them. A colliery at Llanelly absorbed £1,000 of capital in 1613, but this is unlikely to have been typical. Small though they were in comparison with the collieries of the North East, some Midland mines still proved very expensive. It is easy to be misled by evidence dealing only with the physical factors like 'A mapp demonstrating the charge and manner of and for the making of a coale myne in the Hare Close in Strelley', dating from mid-seventeenth century, which estimates the total cost of a mine with three pits and two water-driven pumps at £654 5s. In this case, as the mines were being worked by the freeholder, there was no rent to pay; but when, in the early years of the century, Huntingdon Beaumont had worked Strelley with Wollaton, his operating profit was completely swamped by the exorbitant rent charged by Sir John Byron: £4,000 payable by annual instalments of £500, 'besides usurie, which biteth to the verie boone'.[5]

Again, legal disputes proved extremely expensive to many coalowners, and sometimes employees of the parties came to blows, as at Strelley in 1654. Men employed by one party entered the colliery armed 'with swords and staves' and drove the other party's workmen out. In the previous century a dispute between Nicholas Strelley and Sir John Willoughby went to the Court of Star Chamber. It appears that Strelley sank a pit close to the boundary of the two properties and released a considerable flow of water into the Wollaton workings which he claimed that the Wollaton sough (which

Plate 4   An eighteenth-century drainage adit or sough exposed during opencasting at Newman Spinney, Derbyshire. *(NCB)*

had cost Willoughby a great deal of money to drive) was capable of clearing had Willoughby not deliberately blocked it so as to make the Strelley pit unworkable. In 1545, an agreement was made between Willoughby and Strelley for the latter's workings to be thurled through to the Wollaton sough, which settled the dispute for a time.[6]

In a Leicestershire case of 1694, John Wilkins complained that a sough which he had driven at a cost of £2,000 had been blocked up by his landlord, Beaumont, so drowning out his colliery at Swannington. Beaumont tried to justify his action by accusing Wilkins of monopolising the local supply of coal. There were many similar cases in other coalfields. In the Forest of Dean, with its dozens of free miners, disputes were largely avoided by an ordinance of the Mine Court which made it unlawful to sink a pit close to a sough driven by another free miner.[7]

In some cases, legal disputes caused collieries to stand idle for long periods. Thus, a Leicestershire case of the late nineteenth century between the Snibston Colliery Company (who sought a reduction in royalties to compensate for poor coal and high drainage costs) and their landlords, the Wigston Hospital trustees, cost the company £4,000 in legal charges and laid the colliery and its 400 workers idle in 1882. Even then, the company had to pay the minimum rent of £100 a year and had to keep the workings open and the pumps going which cost them £3,000 to £4,000 a year for no return. According to Thomas Chambers, leader of the Leicestershire Miners' Association, there had been a similar case in 1873. In these cases, the lessees could not afford to work the collieries without a substantial reduction in rent which was denied them. A rather different circumstance applied at Swanwick (Derbyshire) in 1841 when the Children's Employment Sub-Commissioner noted 'England pit, not at work owing to an injunction from the Court of Chancery. Tegg's Pit, ditto.'[8] Sometimes, where the ownership of a mine was in dispute the court would appoint a manager acceptable to both parties to work it pending a settlement.

As against the £6,000 to £7,000 lost by Huntingdon Beaumont and his partners (rather than the £20,000 estimated by Gray) between 1605 and 1612 on their venture in the North East, we may set the £5,000 which Huntingdon's nephew, Thomas, lost on Measham in the ten years from 1611. His loss is simply explained. He paid £500 a year rent for the colliery but receipts never exceeded £400 in a year and were sometimes as low as £10 to £20.[9]

According to a northern mining treatise, published in 1708, boring to prove the coal cost 15s to 20s a fathom, and sinking at least £2 10s or £3 a fathom. Whin (a hard rock) or running sand, or feeders of water could multiply the cost several times over; and the deeper the sinking the greater the cost per fathom. Some shafts were said to cost £1,000 or more, but the average in the North East was nearer £200. At this date, every colliery required at least two outlets, and mines with multiple shafts were common because of ventilation, drainage and transport problems. For drawing water and coal, horse-driven pumps and gins were usual, but waterwheels or windmills were sometimes employed. A whim-gin for a fair-sized colliery required ten horses at £6 to £12 each; and the pump would require the same. Two sledge horses for drawing the coal from the pit top to the coal heap were also necessary. The cost of driving water levels varied widely, and an average can only be guessed at. There were certainly plenty of long soughs by this time costing £1,000 or more. Minor items like ropes, corves and timber would have to be purchased. Marketing and transport costs could be quite heavy. Apart from landsale collieries, most coal was sold in summer, when the coasters sailed to London, and it was necessary to stock up in advance. The author of the *Compleat Collier* suggests £2,000 to £3,000 as a modest provision for this, sufficient for 10,000 to 15,000 tons of coal. Later in the century, much larger stocks were sometimes carried, an example being Houghton Colliery which was said to have had stocks for sale amounting to 159,160 tons at Candlemas (Feb 2) 1766.[10] There may, however, be some error here, since annual output is unlikely to have been more than 15,000 tons.

We may say, then, that an average-sized undertaking in the North-East in the early eighteenth century, would cost at least £1,500 to sink and open out (£1,000 for shafts and water level, £200 for gins, pumps, ropes, etc, and a similar sum for horses, plus, say, £100 for surface buildings). Stocking provisions, wains to carry the coal to the river, and working capital might bring the initial outlay to around £5,000. Difficulty in sinking, or striking faults or other geological impediments, or having a large 'make' of water, could result in this figure being greatly multiplied. The larger undertakings cost between £10,000 and £20,000.

During the course of the eighteenth century, with much of the shallow coal worked out, deeper shafts were sunk. In a Scottish example, a shaft for pumping and a waterwheel cost £1,000 sterling in 1785. Towards the close of the century, there were collieries as deep as 300 feet in Staffordshire and Warwickshire, 420 feet in Yorkshire, 462 feet in Shropshire, 480 feet in South Wales, 500 feet in Somerset, 822 feet in Northumberland and Durham and 993 feet in Cumberland. The expense of sinking a deep shaft can only be justified if a lot of coal is produced from it, so as shafts became progressively deeper, their workings became progressively more extensive. This necessitated improvements in drainage and ventilation. Water levels were still used where possible for drainage, because they cost little to maintain, but increasingly they were supplemented by steam engines to raise water from below the level of free drainage. A Newcomen engine installed in a Scottish colliery in 1725 cost £1,007 11s 4d in materials alone, plus £80 a year for eight years for a licence and the cost of the engine house. As to labour cost, the engineers who erected the engine were also to maintain it and were to be paid £200 a year plus half of the clear profits of the colliery.[11] An atmospheric engine bought for colliery drainage by Boultbee of Swannington in 1760 cost £1,500 and Scottish examples cost about the same. In the early nineteenth century, Farey regarded £2,000 as an average price for a Newcomen pumping engine, as against £500 for a smaller engine used for winding. In 1813, a Scottish colliery owner, Sir John Henderson, planned to spend £3,000 on an engine and pumps for his Fordell Colliery, while an earlier Northumbrian example, installed in 1763 at Walker Colliery, is said to have cost no less than £4,000 to £5,000. The total cost of sinking and opening out this new mine was of the order of £20,000.[12]

Lists of steam engines known to have been in use in certain coalfields have been compiled. Duckham lists seventy-seven for Scotland by 1800, valued conservatively at £120,000. For the West Riding of Yorkshire, Goodchild lists eight atmospheric pumping engines installed before 1750; then forty-five more by 1800. He has also traced one atmospheric winding engine in 1798 and two Boulton and Watt pumping engines. These represent an investment of at least £90,000. For Derbyshire, Nixon lists two atmospheric engines for pumping installed before 1750, ten more plus one atmospheric winding engine and one Boulton and Watt winding engine by 1800, and two more atmospheric engines in 1801 and 1803 respectively. Engines in Derbyshire known to the author, but not on this list, include one at Shipley for pumping dating from about 1750, and another at Pinxton possibly dating from about 1790. These represent an investment of at least £30,000. Further, John Farey, Sen, in the first decade of the nineteenth century, reported having noticed fifty steam winding engines, which he valued at £500 each, in Derbyshire and Nottinghamshire. Smeaton listed a hundred Newcomen engines in the Newcastle-upon-Tyne area in 1769. Indeed, mines of any size in the northern counties had steam pumps by 1800; Whitehaven colliery alone had four. Steam winding engines became general shortly after. Further, many of the northern pumping engines were, like the Walker example, very large ones lifting considerable quantities of water from deep shafts, and being correspondingly expensive. Thus, Percy Main (near North Shields) sunk in 1790 was drained at first by an engine of 63hp and by 1839 required several engines with a combined horse-power of 350. Similarly at Walbottle there were three engines with a combined horse-power of 262. One of the largest engines in this period was erected at Friars Goose about 1819. It was 180hp and drew 1,444,800 gallons of water a day. A few years later there were even larger engines at the new deep pits, for example, one of 300hp at South Hetton.[13]

Besides deep shafts and steam engines, boilers, engine houses and workshops, most northern collieries also had ventilation furnaces by the end of the eighteenth century, and they were served by private railway lines, some of them several miles in length. In 1805, wooden railways cost about £440 a mile (plus the cost of any bridges or tunnels) and iron rail roads cost £900 to £1,000 a mile.[14]

In 1767, there were twenty-four Tyneside collieries with sales of 13,250 tons or more each, and a total sale of 951,350 tons. We would perhaps not be far out if we assumed that the capital invested in these collieries aggregated to £240,000, or something of that order. From this point, the pace of change quickened. John Buddle speaking of the forty-one working collieries on the Tyne estimated the aggregate capital employed by Tyneside colliery owners in 1829 at £1,500,000, exclusive of river craft, and Ross conjectured that the 18 collieries of the Wear represented an investment of almost £1,000,000. Small landsale collieries and others at Blyth and Hartley not covered by Buddle might well have added another £500,000. Ross estimated that another £1,000,000 was invested in new collieries in south Durham between 1829 and 1839.[15] Buddle estimated the cost of sinking and establishing new mines in the coalfield at 'from ten or twelve thousand to one hundred and fifty thousand pounds' and similarly the cost of sinking and establishing new mines in the North East in 1839 was said to range between £10,000 and £150,000, 'including steam engines, railways, staiths, and other appendages'. On the other hand R. W. Brandling in evidence to the 1830 Select Committee, accepted £60,000 as the average cost of 'winning and sinking' a colliery in the district and asserted that some cost above double that figure. Multiplying Brandling's average by 65, the approximate number of collieries then in production, gives a total (£3,900,000) little different from our previous calculation. Further, from evidence given to the Midland Mining Commission of 1843, it may be deduced that the 70 collieries producing then on the Tyne, Tees and Wear absorbed a fixed capital of around £4,336,000 to which a further £500,000 may safely be added for the mines not listed.[16]

Monkwearmouth Colliery (also known as Pemberton Main, which took about eight years to sink commencing in 1826), was said to have cost £80,000 to £100,000 to open out. It was sunk to the record depth of 1,578 feet.[17]

There were large collieries, too, in Cumberland, and some very substantial ones in Scotland. Fordell Colliery, Fife, produced about 30,750 tons in 1791 and one of Lord Dundonald's collieries produced about 32,400 tons in the same year, while the Alloa mines of the Earl of Mar produced over 20,000 tons in 1790 and 48,000 tons in 1810. Govan Colliery, situated only a mile from Glasgow Bridge, had a turnover of over £24,700 in the two years 1804-5. Three seams were exploited at the turn of the century, and steam was used for winding and pumping. The Glasgow Coal Company bought Govan for £30,000 shortly after its formation in 1813.[18] Elsewhere, large mines were exceptional. It is true that some of the South Yorkshire collieries, especially those owned by the Duke of Norfolk, near Sheffield, were expanding from about 1790, but the really substantial expansion did not take place until after 1840.

If we accept John Buddle's estimate of the fixed capital employed on the Tyne and Wear in 1829 as being of the right order, may we extrapolate from this for the industry as a whole? Professor Nef estimated the output of coal in Northumberland and Durham at three million tons a year in the period 1781-90 out of a national total of 10,295,000 tons a year.[19] Assuming that these proportions remained unchanged between 1790 and 1829 and that the relationship between output and capital employed is reasonably constant, the total fixed capital invested in British collieries in 1829 would seem to have been of the order of £10,295,000 exclusive of miners' housing. A similar calculation for 1843 comes to about £16.5 million. However, the mines of the North East were more capital-intensive than those elsewhere, so these estimates are very much on the high side.[20] Nevertheless, they are useful indicators of the order of magnitude.

From about 1840, the pull of rising demand, (whose satisfaction was facilitated by the growth of the railway network) coupled with improvements in equipment, systems of work and management, generated an unprecedented increase in material investment. In 1850, Braithwaite Poole, FRS, estimated that there were 3,000 mines in operation with aggregate capital exceeding £30 million. He calculated that one-third of the national output was produced in Northumberland and Durham.[21]

SOURCES OF CAPITAL

Over the centuries, thousands of pits worked with simple hand tools and windlasses, have been sunk to shallow seams, many of them operated by men of little capital. There was a resurgence of this kind of activity in all old coalfields in the three long disputes of 1893, 1921 and 1926. At Eckington, and no doubt other places, some of these outcrop pits continued for a few years after 1926.

In some cases, yeomen working small pits on their own freeholds were able to expand by ploughing back profits. The Barber Walker partnership may be taken as an example. In the seventeenth century, the Barbers and Fletchers, two interrelated yeomen families, produced outcrop coal, first for their own use and then for the market, on their freehold land at Smalley, Derbyshire. So far, little capital was needed. Towards the close of the century they formed a partnerhsip, and early in the eighteenth century they leased several existing mines belonging to others. The major capital equipment was already there, so they needed only to provide a limited amount of capital which could have come from savings on the working of their older pits. Much of the working capital was provided by butties. The rents paid by the partnership on these early leases were high, sometimes absorbing about a quarter of the total receipts. By the 1730s they were investing substantial sums in soughing, but this work went on side by side with directly productive work, and the coal paid for the development. Their first substantial lease, negotiated in 1738, was for 99 years and the scale of operations gradually increased from this point, although their total sale in

1750 was still a mere £2,500. The increase in coal prices from this time, with fixed rents, was clearly advantageous, as was the advent of canals. By 1808, total output was 67,137 tons and it reached 102,870 tons in the boom year of 1815.

The partnership financed the sinking of a new pit from the profits of existing ones, but some of the capital for expansion was provided by new partners, including the Walker family, substantial farmers at Bilborough, Nottinghamshire, who had worked coal on the land they farmed. When some of the large collieries of the second half of the nineteenth century were sunk, a fresh partnership agreement with separate capital was drawn up for each, the shares held by various partners varying from one partnership to another. However, the whole was managed as one enterprise. Another Derbyshire firm, the Grassmoor Colliery Company, similarly developed from an eighteenth-century mining enterprise carried on by the Barnes family, farmers whose land at Barlow was on a coal outcrop, and who in 1763 leased a pre-existing colliery from the Earl of Oxford's trustees.[22]

One of the most resourceful mining entrepreneurs of the East Midlands, John Wilkins, also came from lowly (probably yeoman) stock. He was a skilled mining man, and owed his good fortune partly to this and partly to his marriage to the heiress of a wealthy family. In the 1780s he took control of a colliery leased by his father-in-law at Swannington and subsequently leased an adjoining mine from the Beaumonts. He spent £2,000 on a sough and was reputed to employ 300 men in 1692. In the 1720s he had two partners. At Swannington he and Captain Adams had equal shares, while at Measham, which he was also now working, Captain Adams had two shares to one held by Wilkins and another by a Mr Sparrow. The annual outputs of Swannington and Measham averaged 7,118 loads and 5,076 loads respectively, a load in this case being equivalent to between two tons and two tons ten cwt.[23] In the period May 1721 to July 1726 Measham made 'profits' (ie net credit balances) totalling £4,664, an annual average of approximately £930; while between February 1724 and September 1727 the deeper and more heavily capitalised Swannington made a total net loss of £60. Measham was still at the stage where new pits were sunk every summer while the existing pits produced vigorously to build up stocks for the autumn when the largest sales occured. Little coal was sold in the winter, but cash flow was maintained by collecting money owed on coal sold by credit in the autumn.

Another interesting feature is that Wilkins had Newcomen pumping engines from Coalbrookdale at both collieries in the 1720s but there is no record of any capital sum having been paid for them. Instead, he paid a rent of £50 a year for each. These engines must have been valued at between £1,000 and £1,500 each at least, so it is unlikely that they were hired from the manufacturers for such a small rental. It is possible that Beaumont bought the engines, leaving Wilkins to pay the annual premium.

Wilkins prospered sufficiently to buy his way into the landed gentry. After his death, Swannington was worked by his former agent, William Newarke in partnership with Christopher Cooke, a maltster, who provided capital. In 1751, they sold the reversion of the lease (10½ years) to a Coventry watchmaker, Gabriel Holland, for £618: £200 to Newarke and £418 to Cooke. The low price should have been a warning to Holland; but he did not heed it. Finding that all the easily-worked coal was approaching exhaustion, he had no option but to sink to deeper seams. Heavy expenditure on sinking and new engines could only be justified by extensive underground working entailing further investment on rails, for example. This was one of the earliest East Midland mines to have rails underground. Holland borrowed from an Ashby lawyer, Isaac Dawson, and incurred heavy liabilities to trade creditors. By the time that all his involuntary developments were completed, his lease was virtually due for renewal and he was unable to raise further funds. The prosperous watchmaker had become an indigent ex-coalowner.[24]

Barber and Fletcher in the Erewash Valley and John Wilkins in Leicestershire prospered during the early eighteenth century when the market for coal, especially in the land-locked Midlands, was flat. Their neighbouring landowning colliery owners were not nearly so enterprising. In 1739, in the Erewash Valley (which straddles the Nottinghamshire-Derbyshire border) most collieries which were not run by Barber and Fletcher stood idle. These included mines owned by the landowners Sir Wolstan Dixey, Sir Robert Sutton, Mr Savile, Mr Plumtre, Sir Charles Sidley, and Sir Windsor Hunlock. They presumably considered that the small profits to be expected did not justify the investment of considerable capital sums for drainage. One prominent landowning family in the area who did invest in Newcomen engines were the Middletons (formerly Willoughbys) of Wollaton, who also owned coal-using enterprises like glass works and iron works. Generally, new sinkings and other capital works were paid for from current revenue, but unusually for estate mines, separate accounts called 'sinking books' were kept for capital items. Also, they did at times borrow heavily from other landowners (eg the Byrons) and from merchants to finance their investments in coal.[25] This family worked their own collieries until the late nineteenth century.

The activities of the Beaumonts of Coleorton are much more difficult to summarise. There were periods when they invested very heavily in their own mines using both estate revenues and borrowed capital; there were periods when, in addition, they leased and invested substantial sums in mines owned by others, and there were times when they let their mines to others. For example, in the 1750s Nicholas Beaumont let Coleorton to Sir Francis Willoughby who paid off his personal debts. Willoughby made profits of £188 in 1576-7 and £248 in the following year from Coleorton. He paid an astonishingly low royalty: 2d per 'rook' (a 'rook' being between 1 ton and 1 ton 10cwt).[26]

The Beaumonts worked their own mines through most of the seventeenth

century, but in the early 1680s both Coleorton and Swannington were
flooded out and rather than invest in costly drainage works and equipment
when the demand for coal was already sluggish, they let them to Wilkins,
whose activities we have already noted. Wilkins concentrated production on
Swannington both as an aid to efficiency, and to depress the total supply of
coal in the area at a time when demand was slack; but this did not suit
Beaumont because he was receiving less in royalties than he had expected
and so he adopted the desperate — and unsuccessful — expedient of
blocking up Wilkins's sough.[27]

The Scottish coalmining industry was dominated by landowners
throughout the eighteenth century. Many worked the mines direct, but
others let them to 'tacksmen'. There were far more mines worked by lessees
in western than in eastern Scotland. Some of the Scottish estate mines were
quite large. For example, Fordell Colliery, Fife, owned by minor lairds
called Henderson, produced about 30,751 tons in 1791, mostly for sea-sale.
In the period 1772 to 1780 inclusive, net profits averaged £1,134 sterling a
year, rising to an average of over £4,000 a year between 1800 and 1808.
Between then and 1812, profits fell substantially and Sir John Henderson
ran up debts totalling almost £9,000. His creditors included the British
Linen Company, bankers (£1,991), his workmen (to whom he owed wages),
another landowner Robert Wemyss (from whom he leased some of his coal),
and sundry trade creditors. Like many Scottish coalowners whose pits were
near the coast, Henderson also worked salt pans.[28]

The Wemyss family were leading colliery owners from the early
seventeenth century, and some heads of the family had considerable mining
and business skill, as their diaries indicate. The extent of their investment in
coal and salt may be gauged from this entry in the diary of the second earl:

> I must show you what these works has been to me since 2nd May 1662 that I
> began [at the harbour or pier] to this 2nd February 1677 being many years.
> The stone harbour was thrice overthrown ere I got it to any perfection and it
> has been to me 40,000 pounds Scots to this day . . . .
> Then the mine for to dry [ie unwater] the seven coals was 30,000 pounds then
> the building of seven pans 20,000 pounds then the big double house and the
> horse work that was five years on coal at the Hill of Methil ere the mine was
> right, cost 10,000 pounds . . . . so do truly declare all was 100,000 pounds . . . .

This is equivalent to £8,333 sterling, a considerable investment for mid-
seventeenth-century Scotland, where capital was notoriously tight.

Generally the Scottish coalmining industry was starved of capital until
well into the nineteenth century and its expansion from 1750 took the form
mainly of a more extensive exploitation of shallow seams. Even in the 1830s,
edge (ie steeply inclined) seams continued to be worked on the bearer system
(where the coal was carried up ladders on the backs of women) which calls for
little capital. On the other hand, while the estate mine remained the typical
unit, there were some substantial commercial concerns from about 1760,

many associated with the rapidly growing iron industry. Of the Carron Iron Company's fixed assets valued at £47,383 in 1769, the mining activities amounted to £11,192. Commercial capital also flowed into the mines around Glasgow from about 1780.[29]

By contrast, the Lowthers are said to have invested £500,000 in only one of their mines at Whitehaven between the middle of the seventeenth century and the middle of the eighteenth century. While there were two other prominent landowning colliery proprietors in Cumberland, the Curwens and the Fletchers, the Lowthers had the lion's share of the Irish market and their collieries were among the largest and most profitable in the country. One, Saltom, wound twenty corves (about 2 ton 10cwt) an hour with a two-horse gin from a depth of 480 feet in 1739. Winding was apparently continuous from Monday morning to Saturday night, giving a a potential annual output from this pit of 18,720 tons. Sir John Clerk estimated the value of a year's output at £4,200 of which £500 to £600 was clear profit. In 1755, there were four Newcomen engines at work at the Whitehaven mines for pumping and in 1776 another was introduced underground.

Gabriel Jars, a Frenchman who visited the district in 1765, said that the Lowthers' mines were then worth £15,000 a year; while it has been estimated that the total output from Whitehaven was about 150,000 tons in 1781. The expansion of the Whitehaven enterprise was a gradual process carried out by skilful men like the Speddings. The very heavy capital investment was financed largely by surpluses on selling coal in a protected market. Competing with the Lowthers in the Irish trade, the Mostyns of Flintshire worked the coal on their estate through five generations and occupied a dominant place in the coal trade of the Dee. However, competition from the efficiently managed Whitehaven collieries and a change in the course of the channel of the Dee resulted in the Mostyn mines being in a run-down state by 1810.[30]

In contrast to Scotland and Cumberland, the North Eastern coalfield attracted much merchant capital at an early date. Some merchant-financiers invested for a short-term gain, but in this the North East was not peculiar. The money was usually lent on the security of the coalmines and if the debt was not discharged on the due date, the financier foreclosed. Huntingdon Beaumont (together with his brother-in-law Sir John Ashburnham) put himself in the hands of four London merchants when he ventured into Northumberland. In 1618, they took possession of Strelley Colliery for the remainder of Beaumont's lease which had five years to run. According to Beaumont, they sold off the stocks of coal amounting to 12,000 tons (worth about £1,800) and then extracted all the easily worked coal and in doing so made it impossible for Beaumont to fulfil the terms of his leases to leave all the pits and soughs in good order. By the time the lease fell in, the merchants had made £9,000 profit on their original investment of £2,000. To recover other debts, they also took over Bilborough Colliery and extracted another

£3,600 profit, leaving the remainder of the reserves drowned. It was another London merchant, John Caris, who disinherited the Strelley family because of their failure to meet a debt of £1,500 negotiated in 1615 on the security of the whole manor.[31]

The Grand Lease (by which the Bishop of Durham's mines in Gateshead and Whickham passed into the hands of a consortium of Newcastle merchants) was negotiated by a London courtier, Thomas Sutton, who is said to have been worth £50,000 on his return to London in 1580. Thereafter, because of the strength of the hostmen's monopoly, London financiers did not play a prominent part in the North East.

Some of the hostmen were of humble origin. One of the most prominent in the early seventeenth century, James Cole, was the grandson of a village blacksmith; another, Adrian Hedworth, had been a quarryman. Other hostmen came from the landed gentry, and others were burgesses of Newcastle of long-standing but, whatever their origins 'the overwhelming majority of them' as Nef says, 'made the money which they invested in mines by setting up, in the first place, as merchants in the towns'.[32]

The amount of capital required to finance a large undertaking could more easily be found by several people than by one, and so partnerships were formed. They also enabled people to spread their investments over a number of separate, though interlocking, firms. This form of organisation was not peculiar to Tyneside, but it was the dominant form there from the middle of the sixteenth century; whereas in most coalfields the entrepreneurial partnership was less important than the estate mine undertaking until the nineteenth century. Partnerships had some of the characteristics of joint-stock companies. For example, their shares were bought and sold. A one-twelfth share of the Grand Lease cost £1,000 in 1587, and yielded an income in 1592 of £150 rising in the years preceding the Civil War, to £260 a year. By contrast, a half-share in a Derbyshire drift mine was sold for £20 in 1587.

The most famous Tyneside partnership was formed in 1726 by Colonel Liddell, the Hon Charles Montague and George Bowes (known as the 'Grand Allies') to exploit large areas of coal much of which lay at a distance from the river. Long railways were built to convey the coals of the 'Grand Alliance' to the staithes. The Grand Alliance was still one of the largest enterprises in 1843 (when it was generally known as Lord Ravensworth and partners). An expert witness reported to the Midland Mining Commission that four undertakings had capitals of the order of £500,000 each: the Grand Allies, Lord Londonderry's trustees, the Countess of Durham's executors, and the Hetton Coal Company. The fact that three of the four largest concerns belonged to landowners should perhaps be taken as a warning that it is unwise to exaggerate the early dominance of the capitalist entrepreneur in this coalfield. The inter-related Quaker families of Pease, Backhouse and Mounsey, whose original capital was drawn largely from the woollen trade, belonged to the second rank of northern colliery proprietors, although people like these also supplied working capital to the landowning colliery owners. Backhouse's Bank was particularly important in this connection.

One part of Northumberland whose development was predominantly undertaken by landowner entrepreneurs was the Blyth-Hartley-Cowpen area where Huntingdon Beaumont had come to grief. The shallow coal here was worked from the thirteenth century mainly for domestic use and for the salt pans, to which the village of Hartley Pans (later known as Seaton Sluice) owed its name. The pits were too far north of the Tyne to share in the early prosperity of that district. In 1670, Sir Ralph Delaval built a harbour at Seaton Burn at a cost of £7,000 to provide an outlet for the local pits, and the Percy family similarly spent over £3,000 on a pier at Cullercoats, two miles to the south, to facilitate the export of coal from Whitley Colliery, in 1677. In the 1760s a deep water dock (the 'New Cut') was built by the Delavals at Seaton Sluice at a cost of £10,000. This widened the demand for coal from the locality, making it profitable to work deeper seams. Much capital was invested in drainage works. By 1770, the Delavals' Hartley Colliery employed 300 workers and produced over 20,000 chaldrons a year.

In North Wales, landowning families like the Mostyns and Wynns were prominent until the late eighteenth century when English capitalists like John Wilkinson, the iron master, became dominant. Also the British Iron Company, whose main centre of activity was South Wales, employed more than 1,500 people in its iron and coal works near Ruabon in North Wales by 1827. In Glamorgan, the Dowlais Iron Co dates from 1763.[33]

The tendency for the size of partnerships to grow, with their parts being subdivided, has been mentioned earlier. To take a late example, John Coke, lord of the manor at Pinxton, Derbyshire, whose family had worked coal there since 1780, took two partners, James Salmond and George Robinson, in 1847 because new capital was needed. At first, the partnership's capital was divided into twenty-eight parts, of which the three partners held twelve, nine and seven respectively; but by 1901, when a limited company was formed, the capital had been subdivided into 1,120ths, all of which were held by descendants of the three original partners. Many of the joint-stock companies formed to undertake coalmining, especially after the passing of the limited liability acts in 1855-62, show the same tendency for the bulk of the shares to be held by a few families. This was so in the case of the New Hucknall Colliery Company (Nottinghamshire) formed in 1877. In 1896, most of the share capital of £111,720 was held by the members of six families. Of these, the most prominent was the Bainbridge family who also had a substantial stake in the larger Bolsover Colliery Company (and several others). Similarly, the Denaby Main Company, formed by Pope and Pearson (an established colliery firm in West Yorkshire) was virtually owned by five men. Of the original share capital of £110,400, J. B. Pope held £16,080, R. Pope £13,140, Joseph Crossley £26,160, George Pearson £16,360 and George Huntriss £18,360. The other shares were mostly held by their relatives.[34]

We have seen how prominent viewers like Mathias Dunn, John Buddle, Jun, and Nicholas Wood became members of colliery partnerships, investing not only their skill but also the money they had saved from their very considerable earnings. For example, in 1843, Westerton Colliery on Teeside was owned by Nicholas Wood and partners. There were similar cases in other districts. One such, Joseph Boultbee, was steward to George Beaumont in the 1750s when Coleorton Colliery was in lease to a local capitalist named Busby. Busby's lease was for twenty-one years, 1738 to 1759 at £150 a year, so he took as much coal as he could out of the pits without ploughing anything back. By the time his lease expired, the 'pit shaft was rotten and the colliery was run in with water'. Beaumont then invited Boultbee to take the lease, and after a year's trial Boultbee agreed. He had to invest £1,500 in a Newcomen engine because the existing pumps were inadequate, and to undertake unremunerative development work. His rent was fixed at £140 a year for 21 years with output restricted to 10,000 loads a year. He also took over Beaumont's Newbold Field Colliery for the same period for £50 a year.

Finding the local market glutted with coal, Boultbee leased an adjacent colliery from Earl Ferrers for £315 a year and virtually closed it down, supplying its customers from Newbold and laying off the officials and workmen. During the twenty-one years of these leases, Boultbee officially sold an annual average of 13,438 stack loads (about 33,600 tons) at prices ranging between 12s and 15s. In 1761, George Beaumont died and was succeeded by his six-year-old son George Beaumont, Jun. Joseph Boultbee's son, Joseph Boultbee, Jun, became steward of the Beaumont estates, and when Boultbee, Sen, died in 1790, took over the collieries too, though on an uncertain tenure.

In 1792, Beaumont employed a surveyor named William Black to investigate the stewardship of Boultbee. Black's report alleged that the original rents were much too low and were so fixed because of the undue influence Boultbee, Sen, had exercised as estate steward; that Coleorton had produced three or four times more than the 10,000 loads a year allowed in the lease; and that Boultbee, Jun, was selling timber from Beaumont's wood and making bricks from Beaumont's clay in considerable quantities without permission. Boultbee's defence was that his salary of £20 a year (the same as his father had been paid in the 1750s) was inadequate. Boultbee was dismissed as steward, and his tenure of the collieries terminated, and a lengthy legal wrangle, which rendered the collieries idle for some years, ensued.[35]

It is easy to understand how Boultbee, Sen, was able to invest in the mines once he had fairly got them going. Paying £1,829 for a new engine for Coleorton in the 1780s was no problem to a man who was, for example, renting a colliery (Newbold) said to be worth £1,000 a year for only £50.

However, he could not have accumulated the initial capital from a salary of £20 a year. Whether he inherited money, or borrowed some, or acquired sufficient in some less reputable fashion, one is unable to say. In a slightly later Nottinghamshire case, a man of humble birth, Thomas North, Sen, apparently accumulated his initial capital from his employment as keeper of one of the district's principal toll-gates.[36] There can be no question but that for a man of little means who wished to become a coalowner, a certain lack of scruple could be a great advantage.

Many, if not most, coalmines in the Midlands and North Wales were run on the charter master or 'big butty' system until well into the nineteenth century. We have already considered the butty's managerial role; here, we are concerned with him as a capitalist. He supplied his pit's working capital: money or credit to pay wages, corves, ropes, timber, candles, powder, ponies or donkeys, provender, sometimes rails and trams, props and tools of various kinds. Usually he was concerned only with the production work of the pit, though in some cases he also looked after shaft maintenance, ventilation and development work. There were some collieries like Swanwick (Derbyshire) where the butties employed even the winding enginemen.

The butty enabled the colliery owner to operate on a smaller capital than he would otherwise have needed. Some entrepreneurs, indeed, would have been unable to carry on but for the fact that the butties supplied the working capital. Many butties owned, or had an interest in, shops or public houses, so that one source of their capital was retail trade and the credit they received from wholesalers. The other chief sources were accumulated savings from wages, and inheritance. Often, a butty took his son into partnership with him; in other cases, the son took over when his father died or retired.[37]

As we have noted, the big butty system lasted longest in South Staffordshire and East Worcestershire where the earls of Dudley were, for centuries, the largest proprietors. Their estate was still 12,000 acres in extent in 1919. In 1690 the mineral property was valued at £150 an acre when coal was sold at 2s 4d a ton. The average price in 1790 had increased to 4s 6d a ton. The thick coal was at shallow depths and there were valuable thin seams and ironstone deposits too. According to John Tryon, in the whole district, weekly output was 46,000 tons in 1817 (at which date there were said to be 2,000 employees), but in 1875 this had risen to 10,500,000 tons a year, of which Dudley's 200 shafts produced 1,500,000 tons. While there was plenty of Staffordshire thick coal left, major capital expenditure was modest. Some collieries were let to other entrpreneurs, at a royalty of about 2s a ton in 1835, falling to 4d a ton in 1919. Others were in the direct possession of the earl, but the circulating capital was provided by the butties. The coal was easily worked, overall productivity being (according to Tryon's figures) about 23 tons per man per week (say 4 tons per shift) in 1817, about four times the long-run national average and well above A. J. Taylor's estimate for mid-century, despite the primitive equipment. (It is likely that Tryon's estimate of the numbers employed was an understatement, since the owners did not know how many employees the butties had at this date. Even so, productivity was certainly high.) Most of the coal was sold within a radius of 20 miles of the collieries. The earl therefore enjoyed a high rate of return on his capital.

Higher efficiency would have produced a more rapid depletion of the shallow, easily-worked reserves, making heavy expenditure on deep shafts, pumps, fans, winding engines, etc, necessary. Until the late nineteenth century his mines operated at an equilibrium where modest inputs of capital were sufficient to maintain a steady, and indeed a rising, income stream. Under these circumstances there was generally no incentive to invest more heavily, although towards the middle of the nineteenth century under the management of Richard Smith, the Dudley mines became far more enterprising for a time. Smith, a mining engineer, replaced F. Downing, an old-style land and mineral agent. Smith abandoned the policy of conserving some of the mines 'as an asset'. He continued to work the thick coal, but let out at royalties as high as 20 per cent the ironstone, thin seam and 'broken' workings (ie workings where only the pillars were left). Between 1836 and 1847, the 'net profit' on the estate's mines and other industrial activities rose from £25,005 to £157,990, of which £74,955 came from royalties paid by leaseholders. Between March 1845 and December 1859, 'net profits' totalled £1,542,391, and Smith received over £22,000 in commission.[38]

Some of the lessees spent a fair amount on improving shafts and modernisation; but the mines worked on the butty system which were not let, were still small and inefficient in 1907. By this date, the earl himself had one large new mine, Baggeridge, sunk to a depth of 550 yards at a total cost of £400,000 by a company (The Baggeridge Colliery Ltd) of which he was 'almost the sole shareholder'. Dudley had to borrow half of the capital, and up to 1919 all the profits and royalties had been ploughed back to pay for development. This was in marked contrast to his experience with the shallow mines of an earlier period. Dudley also had a modern steel works (again owned by a limited liability company of which he was virtually the sole shareholder), brickworks, lime kilns, and an aluminium foundry, which provided a protected market for much of his coal.

A. J. Taylor has estimated that there were nearly 400 collieries in the Black Country in the middle of the nineteenth century, and that their average annual output was 15,000 tons with a labour force of sixty to seventy. He says that the average colliery 'represented a capital commitment of no more than £3,000'. Presumably, this relates to the fixed capital only, since the butties provided the working capital. Similarly, a witness told the Midland Mining Commission that the average cost 'in sinking and engines', of a South Staffordshire colliery, was £3,000 to £4,000, with some costing 'as little as £1,000'.[39]

In the second half of the nineteenth century, the limited liability company became the dominant form of mining undertaking, with partnerships like Barber Walker becoming private limited companies. As we have noted, a considerable part of the equity of even the public limited companies was subscribed by a comparatively small number of families, most of whom had been associated with mining for generations. It may be that people with no knowledge of mining were unwilling to risk their capital in a form of investment whose rewards were uncertain.

Colliery undertakings borrowed from banks and other institutions. For example, Wrights, the Nottingham bankers, provided Thomas North not only with working capital but with long-term capital, too. When he died in 1868, he owed the bank about £200,000. As security for the loans, North had bequeathed to Ichabod Charles Wright all his colliery and ancillary property 'in lieu of payments of principal and interest'. North also borrowed money from the Union Loan Club.[40]

It was not easy for small savers to find investment opportunities which were both remunerative and sound. Loan clubs, which channelled small savings to industrialists and others needing capital, performed a useful function. Occasionally, these and similar institutions made the mistake, however, of putting too much money into one firm. The Sheffield and South Yorkshire Building Society, for example, invested £117,885 in Dunraven Colliery debenture stock, and because of the colliery's inability to either pay principal or interest, had to take it over. In 1890, this colliery changed hands for only £20,000 but in the meantime, the building society had had to go into liquidation.[41]

In order to avoid going over the same ground twice, limited liability companies will be considered further under the next section on profits.

PROFITS

Nef's observation on profits, that:

> The successes in the Newcastle district, during the reigns of Elizabeth and James I, were due in large measure to an exceptional combination of favourable circumstances, which did not occur again until the latter part of the eighteenth century, when, with the general adoption of coal for smelting iron, the consumption in manufactures increased so rapidly that demand once more tended to outrun supply

can be accepted as a broad general statement.[42] The prosperity of the earlier period was shared by most districts. In the second half of the eighteenth century, the growth in the demand for coal was accompanied by improved means of transport, and here we should not underestimate the value of turnpike roads to land-locked districts like Leicestershire and Warwickshire.

In an industry as diverse as coalmining, the profit records of different undertakings varied widely. Even in the most profitable period, some collieries made losses; and in the most depressed period some collieries made profits. Runs of good years and runs of bad years were common. These could be due, among other things, to persistently favourable or unfavourable geological conditions; to good or bad management; or to strong or weak marketing conditions. It is impossible to estimate an average profit for any period earlier than the late nineteenth century. The undertakings whose records survive are unlikely to have been typical. Also, early accounts are difficult to interpret. Before mid-nineteenth century, capital and revenue items were usually lumped together, and the failure to allow for depreciation, or the depletion of mineral reserves, could give an altogether misleading impression of profitability. One of the few landowning colliery owners to recognise this was Earl Fitzwilliam. From 1836, the Fitzwilliam collieries were debited with the value of coal extracted at between £120 and £300 an acre, depending on the seam. This was a sensible way of making financial provision for the replacement of a wasting asset.

The Fitzwilliam estate was one of the largest enterprises of its kind. In mid-nineteenth century, it had two groups of collieries with a combined annual average output of over 300,000 tons. Between 1798 and 1856 they sold coal valued at £2,123,110 while expenses charged to the colliery accounts aggregated to £1,950,273, giving a surplus of £172,837. However, included in the expenses were interest on Fitzwilliam's own capital employed in the business totalling £129,027, and debits for Fitzwilliam's own coal got totalling £112,436. Adding these to the surplus, we arrive at a total contribution of £414,300. To arrive at a net contribution to the Fitzwilliam treasury, Dr Mee has deducted the notional value of the collieries' capital (£100,000) since this appears to represent land purchased to augment reserves of coal, plus miscellaneous items bought for colliery use on the 'Household General' account.[43]

The Wollaton mines were producing an annual profit of over £200 a year by the end of the fifteenth century and they remained generally profitable thereafter. Throughout the sixteenth century, profits from Wollaton of several hundred pounds a year were recorded, and in addition there were profits on a similar scale from other collieries worked by the Willoughbys.[44] By the second half of the eighteenth century, annual profits of around £6,000 were being produced from Wollaton and Trowell Field. Towards the century's close, these mines were being challenged by the nearby Shipley collieries owned by the Miller-Mundy family which grew as a result of the building of canals. In the period December 1803 to December 1810, Shipley's sales varied between 25,000 and 49,000 tons a year, the average annual value being £17,836, and the average annual surplus, £5,898.[45]

The opening up of the South Wales coalfield owed much to another landowning family, the Butes, who were also active in Durham. John, Marquis of Bute, said in 1919 that his grandfather was the first to sink to the steam coal in the principality. This was the Bute Merthyr Colliery at Treherbert opened in 1850. They also built Bute Dock (commencing in

1830) at a total cost of £5.5 million on which they received interest of only 1½ per cent per annum. They had always worked some of their own minerals, but most were let. In 1919, they were still working one colliery, but only because the original entrepreneurs, after losing £20,000 to £30,000 on it, surrendered their lease, and his father then worked it for some years at a loss until it started to pay.[46]

The profits of Measham and Swannington have been mentioned earlier. Another mineral estate in the same part of Leicestershire was Staunton Harold, owned by the Ferrers family. The main colliery here was Lount, whose accounts for 1763 to 1776 survive. There was an excess of income over expenditure in seven of the fourteen years, but the 'losses' exceeded the 'profits', chiefly because of the results for 1769 when income was £1,170 against expenditure of £1,661. The highest surplus in any one year was £181 for 1775, and the highest deficit for any one year, apart from 1769, was £238 for 1765. As with most estate mines, no distinction was made between capital and revenue expenditure. The earls Ferrers owned brick works and lime-kilns, both of which benefitted from cheap coal. The family continued to work some of their own mines until mid-nineteenth century, and there is no doubt that they made a considerable profit. For some of the time, there was little competition. Beaumont's mines were out of production for twenty-seven years from 1793; Godolphin Burslem's Thringstone Colliery nearby ran into a major fault in 1796 and was abandoned; while Fenton and Raper's Swannington Colliery (leased from Wigston Hospital) closed in 1800, and was not replaced until 1805. In 1811, when his coal was selling at £1 a load against the 6s to 8s a load of the 1790s Ferrer's Lount Colliery was said to have 'lately become very extensive', while his new colliery at Staunton Harold was being 'worked with considerable effect' in 1802.[47]

The mining partnership, which had long been the predominant form of colliery enterprise in Northumberland and Durham, was becoming increasingly important in the Midlands. Some of these partnerships were quite large, but size was no guarantee of success. Perhaps we may illustrate this by taking two examples from the same district: the Erewash Valley. One, Benjamin Outram and Company, founded in 1792, and later re-styled the Butterley Company, was highly successful; the other, North, Wakefield and Morley, was a financial disaster.

The four original partners in Butterley were Benjamin Outram and William Jessop, engineers, Francis Beresford, gentleman, and Beresford's son-in-law John Wright, member of a prominent Nottingham ironmongering and banking firm. Each partner was supposed to pay £6,000 as his share of the initial capital, but a memorandum of 1798 records that of the nominal capital at that date of £32,000, Wright's contribution was £14,000 against £8,000 from Beresford, and £5,000 each from the other two. By 1858, when a two-fifteenths share changed hands at £58,000, the nominal capital had risen to £436,000. A one-fifth share changed hands in

1883 for £100,000 and when the firm adopted limited liability in 1888, its initial share and debenture capital was about £740,000. At 1 January 1947, the issued capital was £2.3 million and £1 shares were quoted at 53s. The policy of the company of limiting dividends so as to ensure continuous growth was possible because almost half the shares were still held by the Wright family. The last colliery opened by Butterley was Ollerton, sunk between 1921 and 1927 and not fully developed until 1937, which cost £840,566 to sink and equip. The building of New Ollerton village brought the company's investment there up to almost £1.5 million.[48]

North, Wakefield and Morley developed out of a small family firm started by Thomas North, Sen, a toll-house gate-keeper and coal merchant, who started to work coalmines in the 1820s. By 1846, there were four partners Thomas North, Jun, Thomas Wakefield, James Morley and Samuel Parsons, Jun, and the capital was assessed at £80,312. The partnership sank the first recognisably modern colliery in the district in 1841-3 and other sinkings followed. In 1847 Wakefield went bankrupt, chiefly because of his mining venture, and shortly afterwards, North became sole proprietor of the firm, but he was heavily indebted to Wrights, the Nottingham bankers, whose connection with Butterley we have noted. When he died in 1868, he owed them £224,000 and they took over his business property. Had North lived a few years more, the boom of 1871-3 would have enabled him to clear his debt and to amass a surplus, because by that time his collieries had a capacity of at least 250,000 tons of coal a year which they could have produced and sold at a high level of efficiency.[49]

The marginal return on extra output in a boom period was high: sufficiently so to enable many of the high-cost producers to recoup the losses of lean years. Further, new firms whose collieries were just coming into production at the beginning of a boom, had a double bonus because costs of production are at their lowest when the workings are close to the pit bottom. The South Normanton Colliery Co Ltd, Derbyshire, may be taken as an example. This was a small company formed in 1892 by a group of Durham people with previous mining interests led by Mr A. Mein of Bishop Auckland. The colliery was sunk in 1892-3, and its winding capacity when fully developed was 500 tons a day. The initial cost of opening out the colliery was £9,829 and investment in fixed plant and machinery to February 1902 was £6,990, with a further £2,712 in the succeeding 12 months. The Boer War boom occurred while the workings were still near the pit bottom, and in one half year alone (to 28 February 1901) a net profit of £7,135 was struck after making an exceptional allowance of £3,500 for depreciation. Ordinary shareholders received a dividend of 50 per cent for the half year, and another 10 per cent for the following half year, while the company's reserves still exceeded the actual cost of material investment.[50]

South Normanton was not the only Midland colliery company to be financed mainly by people from Northumberland and Durham. George

Stephenson's ventures at Clay Cross, Derbyshire and Snibston, Leicestershire are two obvious examples. When they were sunk they were the most important collieries in the East Midlands. By 1820, the colliery owners of the North-East were meeting increasing competition in markets where they had been pre-eminent, while much of the easily-worked coal was exhausted. There was, therefore, a migration of capital to other districts. Pease and Partners, for example, invested in a new colliery (Thorne) in South Yorkshire, while another new Yorkshire company founded in 1900 took its name from Wallsend Main, indicating the origin of the founders.

In the North East itself, the centre of gravity was shifting away from the outcrop. Until about 1820, colliery owners were reluctant to sink shafts through the magnesian limestone of East Durham because they thought it 'cut off the coal measures; or that, if coal existed beneath it, such coal must be deteriorated both in quality and thickness'. Hetton Colliery, one of the earliest and largest mines in the district, had three pairs of shafts which were very costly to sink. The owners, The Hetton Colliery Company, engaged George Stephenson to construct a railway to Sunderland, some six or seven miles distant, and this opened on 18 November 1822. The colliery had a capacity of 530,000 tons a year and according to Galloway 'It proved most prosperous and profitable; indeed, the profits are characterized as extravagant'. In its first year, the company is said to have made a clear profit of £80,000. Judging from the prices at which the shares changed hands, the company was valued at £324,000 rising in a few years to £500,000.[51]

By contrast, the Durham County Coal Company, which owned Byers Green, Evenwood, Coxhoe and Whitworth Park Collieries, was unprofitable. These collieries were linked to the Tees by the Clarence Railway. Whitworth Park, which opened in July 1841, was leased from Mr R. E. D. Shaftoe whose ancestors were hostmen working their own collieries,[52] but he preferred the certain income from royalties to uncertain profits. The company's outlay on this colliery was nearly £40,000 and operations commenced during a slump when the return was poor. In 1842, the colliery was closed and much of the equipment dismantled. According to Ross 'many of the original shareholders have retired with loss, and . . . the general prospects of the company are still inauspicious'.[53]

There were comparatively few joint-stock companies in coalmining until after the limited liability Act of 1862. Many of the old partnerships then became private limited liability companies, but the total number of shareholders remained small. Indeed, even the public companies (about half the total in 1919) attracted investments from quite a narrow range of investors, many with family mining connections. Even in 1919, there were only 37,316 separate holdings in purely coalmining firms, and 94,723 in firms with both coalmining and other interests. The number of individuals holding shares was very much lower than this, since most investors held shares in many different companies.[54]

Plate 5 An iron pit-tub found in early nineteenth-century workings, 60ft deep, during opencasting at Blaenavon, South Wales, *(Western Mail and Echo,* Cardiff).

The joint-stock companies established in the second half of the nineteenth century had varying fortunes. Many of them invested heavily during booms (especially 1871-3 and 1900-1) only to have the new capacity ready to produce after the boom had collapsed. One of the largest of these firms was the Powell Duffryn Steam Coal Co Ltd, formed in 1864, which employed over 18,000 men (not all of whom were miners) and produced nearly four million tons of coal a year by 1919.

Between 1873 and 1888, Powell Duffryn paid no dividends in cash, but paid 9.77 per cent in shares for the whole 15 years. By 1888, the company was bankrupt, and Joseph Shaw, a director, advised the board to wind it up. It was as well that this advice was ignored, because improving trade from 1888 enabled them to profit from their massive investments. A dividend of 3 per cent was paid on the ordinary shares in 1889, followed by 12½ per cent and 7½ per cent in the two succeeding years. In four of the following seven years only 1½ per cent was paid, and in 1898, there was no distribution. Over the forty years 1873 to 1913, Powell Duffryn's shareholders received £2,144,301 in cash and £594,757 in shares, giving an average return of 7.36 per cent less tax. In the same period, the company's wages bill totalled £30 million. Between 1898 and 1918, Powell Duffryn invested £1,731,157 in new sinkings, coke ovens, coal preparation plants and the like.

New companies in South Wales which did particularly badly during the 'Great Depression' of 1873-96 included Rhondda Merthyr, which was abandoned after losing £80,000 to £85,000 in ten years; and Gueret's Craigola, similarly abandoned after losing £105,000 in ten years' working. Another example, Naval Colliery, cost its founder firm between £100,000 and £150,000; they then gave up and another company was formed by nine people who put £60,000 into it. After working the colliery at a loss for some years and losing both its original capital and a further £50,000 or so of working capital, this company collapsed and its assets were taken over by its principal creditor, a coal exporting firm. In 1919 the chairman of this reconstituted Naval Colliery Co Ltd was T. J. Callaghan who was also chairman of the parent company L. Gueret Ltd and of the South Wales Coal Exporters' Association.[55]

Even the old established north-eastern firm Pease and Partners, now a limited company, was only able to pay its shareholders an average of 1.6 per cent per annum between 1882 and 1892, though from 1904 to 1913 it paid an average of just under 11 per cent. In one period of five years, it paid nothing at all. Commenting on this, Baron Gainford said 'It is the speculative character of the Industry which attracts the private investor.' It would perhaps have been clearer to say that it was the prospect of great gain which induced the investor to put his money into anything so risky as coalmining.[56]

Another example quoted by Lord Gainford was Horden Collieries Ltd, the establishment of which in 1900 was criticised by the financial press as 'highly speculative'. This company purchased three colliery firms, one of which had never paid a dividend, while the others had paid 2 per cent and 5 per cent respectively. They 'all came to grief owing to water trouble and the whole of the share capital amounting to about half a million pounds, spent over nearly 25 years was lost.' The public only took up £45,000 worth of shares in the new venture and the remainder of the original share capital of £250,000 was subscribed by the directors and their friends. Great technical difficulties were encountered in sinking new shafts and unwatering the old, and in its first seven years the company spent over £800,000 without paying a penny in dividends. By 1919, the issued share capital had increased to £983,310 and there were debentures of £300,000 and other loan stock of £139,000 outstanding. The mines were then producing 6,000 tons a day and had the capacity to double that when they were fully developed. In 1920, one bonus share was issued for every two held; and in 1924 a dividend of 12½ per cent was declared.[57]

The Shireoaks Colliery Co Ltd is another interesting case. This is in North Nottinghamshire but is normally regarded as belonging to South Yorkshire. The concealed coal measures in that locality were proved by J. T. Woodhouse in 1839 and it was, indeed, on the basis of his findings there that he recommended Thomas North to sink a colliery at Cinderhill, some 22 miles to the south.

The owner of the Shireoaks royalty, the Duke of Newcastle, financed the sinking of Shireoaks to the Top Hard Seam in 1854 and appointed Mr John Lancaster as engineer in charge. The shafts were 515 yards deep and took about five years to complete. This was considered to be a very risky venture, and the Duke of Newcastle was awarded a medal at the Great Exhibition of

1862 for undertaking it, thus opening up the concealed coalfield of South Yorkshire. Other owners benefited from Newcastle's venture more than he did. Once he proved the concealed coalfield as far east as Shireoaks, it was obvious that the coal measures were continuous from the exposed coalfield in West Yorkshire and Derbyshire.

The Shireoaks Company was established in 1864 and took possession of the colliery on 1 January 1865 when capacity was about 180,000 tons a year. They agreed to pay the Duke of Newcastle £100,000 for the firm as a going concern, although he retained the freehold as was usual. Operating profits in the early years were barely satisfactory:

1 April to 31 December 1864	£10,679
1 January to 30 June 1868	£5,983
1 July to 31 December 1865	£5,242

Recommending the sinking of a new colliery at Southgate to the west of Shireoaks in a report dated 30 January 1867, the manager, Mr C. Tylden Wright pointed out that the benefit of the Shireoaks development had been reaped by:

> Other parties . . . by leasing and working coal proved to exist to the west of this colliery at a much smaller depth, and as these concerns are brought into full operation at less than one-quarter of the cost at Shireoaks, they will always be able to sell Coal at such prices as will leave them a fair profit, but would yield a very small dividend on our large capital.

It was estimated that the cost of sinking two 13ft diameter shafts at Southgate to the Top Hard Seam, 246 yards deep, would be about £15,700.[58]

Turning now from the particular to the general, in 1919 there were about 3,300 collieries with 1,452 coalowners, 434 of whom produced less than 2,000 tons of coal a year each. In the five years to 31 December 1913, output averaged 270 million tons a year which sold at an average pit-head price of 8s 9d a ton. Profits after depreciation (but before deducting interest or royalties) averaged £19 million. Of that, royalties absorbed £6 million, leaving £13 million equal to $11\frac{1}{2}d$ a ton.[59]

J. C. Stamp suggested that from the £13 million, it was necessary to deduct between £1 million and £2 million to take account of the fact that coal is a wasting asset. The net return on capital in coalmining before World War I (estimated at £135 million) was thus rather less than 9 per cent, compared with 9 or 10 per cent in industry generally, whereas because of the high risk, a return 2 or 3 per cent higher than the average could have been expected. This may appear to bear out the assertion made by long-established coalowners that too much new capital had entered the industry during the booms of the later nineteenth century, causing overproduction. The excess of capacity was, however, merely a short-term feature felt particularly in the slumps of 1874-9, 1883-7 and 1893-6. In most cases the new mines sunk to exploit virgin reserves in, for example, the concealed coalfields of the North East, Yorkshire and Nottinghamshire made high profits on the upswing of the trade cycle, and at least broke even on the downswing. It was the older enterprises, working high-cost coal, which suffered in the main.

Just how wide were the variations in profit between one colliery owner and another was shown in a study of November and December 1917. Some 675 undertakings, comprising 46 per cent of the total number (and producing three-quarters of the total output) were taken. Thirty-one per cent of these undertakings produced 62 per cent of the total output at an average profit of 2s 3d per ton; while 15 per cent of the undertakings included in the study produced 13 per cent of the total output at an average loss of 2s a ton.

Some individual companies made profits as high as 6s a ton, while others lost a similar amount. Mr A. L. (later Sir Arthur) Dickinson concluded: 'It is perfectly clear that the price of coal that is a fortune for some collieries spells bankruptcy for a number of others.' Many high-cost producers remained in business, working at a loss during slumps, in the expectation of profit once trade improved and it was their low average rate of return which pulled down the general average for the industry.

The people with assured income from coal were the royalty owners many of whom were landowners who had once adventured their capital in mining, but who now preferred to leave the risks to capitalist entrepreneurs. It was not uncommon for such people to hold shares in the limited companies which now worked their coal but this entailed a minimum of risk and none of the worry of management, although it helps to explain why they were prepared to accept royalties which were much lower, relative to the value of the coal, than at an earlier period. In 1918, there were 3,789 royalty owners who received in total £5,960,365. At one end of the scale, ninety-seven received less than £5 each while at the other end of the scale, seven received over £79,000 each. The Church had by far the largest royalty income: about £410,000 in 1917 and £423,448 in 1918.[60]

VERTICAL INTEGRATION

Coalmines and their associated railways occupy a fair amount of land and, until recently, employed many horses. Also, colliery owners needed a stock of land for future tipping of spoil. All colliery undertakings of any size, whether estate mines or capitalist enterprises, therefore had farms attached to them.

Colliery owners found it difficult to sell small coal, so they used as much as they could themselves. In coastal coalfields, almost all early major coalowners owned, or had a share in, salt pans, while in many areas they had lime-kilns. The original name of the company which sank Sutton Colliery, Nottinghamshire, in 1873 (The Skegby Colliery Brick and Lime Company) indicates another activity closely connected with coalmining: the manufacture of bricks, tiles and other ceramics. Very often, clay suitable for firing is found in association with coal, and on the other side of the coin, colliery firms used a great many bricks both about the mines and for constructing workmen's cottages.

Similarly, ironstone is found in most coalfields and after Abraham Darby's successful use of coke for smelting in 1708-9 undertakings which mined both coal and iron often established their own furnaces. Of course, some landed proprietors like the Willoughbys of Wollaton and the Zouche family of Codnor had iron furnaces and forges long before this; but the great era of the integrated undertaking, producing and using coal, lime and ironstone and in many cases fabricating the finished product, began in the second half of the eighteenth century. The largest of these was the interconnected Staveley and Sheepbridge enterprise of the East Midlands, but there were sizable undertakings in every district. Also, most districts had examples of similar indifferently successful enterprises like the Dalmellington Iron Company of Ayrshire which smelted iron from 1848 until 1921 and on the whole made very little profit out of it. Subsequently, it concentrated on the production of coal.[61] From the 1780s the manufacturers of wrought iron, like the Butterley Company, faced increasing competition from steel made mainly with imported ore. Also many British ironstone deposits fell below the margin of production in competition with the richer imported ores.

One Welsh firm the Dowlais Iron Company, owed its early profitability to a 99 year lease granted in 1763 of over 10,000 acres of manorial waste in Glamorgan at a very low rental.

Besides producing coke for iron furnaces, many colliery companies also had small gas works to supply light and heat to the mine and the mining village, and gas coke was a residual. After the inventions of Coppée (1861) Carvés (1866) and Simon (1881) of coke plant to recover by-products, some colliery companies turned their attention to the manufacture of dyes and other chemicals which had previously been a virtual monopoly of Germany, and in the twentieth century a few also manufactured petroleum products from coal.[62]

Some vertically integrated companies used a high proportion of their own coal which gave them some protection against fluctuations in the demand for domestic qualities. When the demand for iron products fell flat, the integrated coal and iron companies then flooded the domestic coal market, thus depressing prices. Also, by charging themselves artificially low transfer prices for the coal they consumed they could depress the profit figure of their collieries and since miners' wages were tied in one way or another to prices and profits, the miners' unions in the twentieth century viewed such concerns with suspicion.

3

Labour

THE STATUS OF THE LABOURER

It was suggested in Chapter 1 that some miners in the Middle Ages were serfs but there was an element of speculation in this. The unfree status of Scottish pitmen at a later date can however be asserted as a fact.

Serfdom in Scotland was not a survival from medieval times, but a deliberate operation of the Scottish Parliament dating from an Act of 1606. This, and subsequent legislation, were somewhat obscure. However, as Duckham tells us:

> Broadly speaking, estate custom and legal decisions established that colliery labourers were bound to the coal they worked; that they became so bound by either a formal pact or (more likely) by uninterrupted labour in a mine for a year and a day; that effective desertion for such a period was a presumptive ground for freedom; and that as serfs they could be removed to other mines of their master, or be disposed of by sale or lease with a 'going coal-work'.

The giving and receiving of 'arles' (binding money) was usually regarded as an essential element in the process; and many coalowners gave arles at the baptism of children born to their serfs.[1]

Collier serfs were not slaves; they could own and bequeath property and undertake legal actions. Some became colliery officials and a few leased small mines. Most oversmen were serfs: and the diaries of the earls of Wemyss show that serfs were also appointed as grieves in some cases. Two such were James Black and James Lumsden, who were grieves in 1677.[2]

Servile labour was not cheap. According to Duckham, shortly after the abolition of serfdom in 1809, 'the hewing rates of coal were 2s 11d to 3s 4d a ton in Midlothian, compared to only 1s 6d per ton in Yorkshire and 1s 1d per ton near Newcastle'. On the other hand, the English collier had a considerably higher average productivity. Also, Scottish colliers paid their own bearers (usually their wives and daughters) out of their tonnage rates. It seems likely that the average earnings of Scottish miners (taking bearers into account) were lower than those of English miners from the late seventeenth century to the early nineteenth century, but that unit wages costs in Scotland were greater than in many English mines because of a generally lower productivity.[3]

Duckham suggests that the purpose of serfdom was to ensure a sufficient supply of labour, and that its abolition in the late eighteenth century was due to the fact that serfdom had become a positive bar to recruitment. The expansion in the demand for coal necessitated the recruitment of people from other occupations who were unwilling to accept servile status. There were two emancipation acts, one in 1775 which provided for enfranchisement by stages and another in 1799 which completed the process.[4]

From at least the early eighteenth century until 1844, miners in Durham and Northumberland were commonly bound for a year at a time. The terms of employment were set out in a document called a bond, to which all the workmen subscribed their signatures, or marks. A typical bond dated 8 October 1788 between John Nesham, Esq of Houghton-le-Spring and the 105 workmen at his colliery at Newbottle Burn Moor, County Durham, is exhibited at the Lound Hall Mining Museum. There were 59 hewers, 12 barrowmen, 12 horse drivers, 12 lampkeepers and 10 deputies. Only 16 were able to write their names: 11 hewers, two horse drivers and one each of the other classes. The prices for the various classes of work were specified and wages were to be paid at 28-day intervals. Hewing, heading and putting (ie haulage) were paid by piecework. Hewing, for example, was paid for at the rate of 3s or 3s 3d per score. The size of the score varied from pit to pit, but in this case was 21 corves of 20 pecks (approximately four tons). The horse drivers were paid 1s a day. Men who were absent from work without good cause had to forfeit 1s a day, but on the other hand, for every day they were idle through some fault of John Nesham's, they were entitled to receive 1s from him. The average daily output of a hewer was probably at least two tons giving piecework earnings of 1s 6d or so a day, plus payment for certain tasks other than hewing which probably brought the average daily earnings to around 2s.

Hewers at Mr George Humble's Lemonton Colliery (Durham) were paid 1s 8d per score in 1749, but here 12-peck corves were in use, so the tonnage price was about 8d compared with about 9d at Newbottle. In another Durham example of 1751 (Morton Hill) the hewing price for 14-peck corves was 1s 10d per score, again about 8d a ton.[5]

At the time of hiring, each man was given a sum of binding money. At Byker Hill, Northumberland, in 1774, this was only 1s. However, when trade was particularly good, the competition for labour forced up the level of binding money. By about 1800, the usual amount had risen to 2 or 3 guineas, but in 1804, 12 or 14 guineas were given at Tyneside collieries and 18 guineas on the Wear. In 1805, the owners resolved that the binding money for a hewer who was a householder should not exceed 3 guineas on the Tyne and 5 guineas on the Wear, with slightly higher amounts for single hewers (£3 13s 6d on the Tyne and 6 guineas on the Wear) and smaller amounts for other classes of labour. Pressure from the men resulted in payments of 5 guineas on the Tyne in 1809 and 1810. The 1805 arrangement made by the owners' association also provided that there should be no advance on the wages for hewing and haulage work. Such understandings were in breach of the

Combination Acts.[6]

For a bound miner to abscond was a crime punishable by a fine or imprisonment, as well as a breach of contract. Further, owners sometimes resolved not to engage a man previously bound to another owner so he had little option but to renew his bondage. One of the owners' resolutions of 1805 was that no owner should engage more than ten men or twelve boys from another colliery. The one advantage of the system to miners was that it guaranteed them a wage, although at a reduced rate, when the pit was idle. In 1826, the guaranteed wage for hewers was about 2s 6d a day, but in 1830 many owners withdrew the guaranteed wage and average earnings fell as low as 8s to 10s a week at many collieries. In 1832, the bond was so worded as to give the impression that there was a guaranteed minimum of 28s per fortnight, but one of its clauses provided a loophole for unscrupulous owners to pay a lesser sum or nothing at all in some circumstances. Agitation against the evils of the bonding system intensified and a strike in 1844 virtually brought it to an end. The man had demanded six-monthly bonds giving a guaranteed five-day week or 15s in money, but the owners preferred to abandon the system.[7]

Hiring for a year or so at a time had apparently been common in many coalfields, and traces of the system were found by the Children's Employment Commission in 1842 in parts of Scotland (where it had replaced serfdom), Lancashire and South Derbyshire. An isolated Nottinghamshire example of a hiring for six months, with 'a sovereign' as binding money, is dated 1848, but by this time contracts terminable at two or four weeks' notice were almost universal.[8]

Whatever the pitman's formal status, he seems to have been regarded with distaste or worse in every district. This may have been due in part to the reality of his servitude. Again, when personal cleanliness was not esteemed very highly, the homes, clothes and skin of miners were permanently discoloured. In new mining districts, the invasion of dark-skinned strangers housed in mean hovels, was inevitably unpopular and miners were regarded with hostility by the original inhabitants.[9]

In rural districts where mining remained closely allied with agriculture and in urban districts where it was only one of several possible industrial occupations, there were no bars to intermarriage between mining and non-mining families. However, pitmen living in the isolated pit villages of Durham, Northumberland and Scotland were regarded as a separate race, and intermarriage was rare until well into the nineteenth century. The Methodist revival which was particularly successful in colliery districts, did much to improve the coalminer's image.[10] In the second half of the century, the rapidly rising demand for coal necessitated an increase in the labour force which could only be met by attracting men away from other industries and services, and so the wages of coalminers rose relative to those of alternative occupations. High wages alone did not improve occupational status, but they

certainly helped, by sustaining a higher standard of living. In the East
Midlands, for example, coalmining became more highly regarded as an
occupation than either agriculture or hosiery. And as the proportion of
miners to the labouring population increased, spread over a much larger
number of families and over wider areas of the country, so the status of
coalmining improved.

NUMBERS EMPLOYED

Before the sixteenth century, there were few full-time coalminers, and even
in the nineteenth there were still areas where many worked for part of their
time on the land.[11] But by the eighteenth century, the typical colliery in
Durham and Northumberland had a hundred or more employees. In
landsale districts, where the market area of a colliery was restricted by heavy
transport costs, the typical unit remained small, but in districts near the sea
or a navigable river, the typical colliery (like Newbottle Burn Moor) was able
to sustain a large labour force. In no other industry at this date, isolated
examples apart, were such large numbers brought together in one
undertaking.

In the nineteenth century, with increasing demand made effective by
canals and railways, and improved mining techniques, collieries employing
500 or more people became common. For example, there were 502 employed
at Jarrow and 857 in the two pits of Haswell Colliery in 1843. By the
century's end there were plenty of mines with over a thousand employees.

The total numbers employed are perhaps easier to estimate for coalmining
than for many industries. This is particularly true of the North East coalfield
where the size of the labour force can be inferred from the output estimates.
Nef estimates the numbers of miners in the Tyne Valley alone at 3,000 in
1637. For the end of the seventeenth century, his estimate is 3,000 to 4,000
for the Tyne, 1,500 to 2,000 for the rest of Durham and Northumberland,
3,000 for Scotland and 15,000 to 18,000 for Great Britain as a whole. In
contrast to the Tyne, the Trent Valley, which was one of the busiest of the
inland coalfields, probably employed between 150 and 330 at the beginning
of the seventeenth century, and between 500 and 1,000 at its end. By 1800
coal output in Great Britain was about 10 million tons a year, and assuming
an output per man-year of 200 tons, the labour force would then be around
50,000. The assumption of an output per man-year of 200 tons is based on
the fact that productivity in coalmining tends to average out at somewhere
around a ton per man per shift. With bell-pit working, outputs of about
15cwt to a ton per man-shift can be expected. As mining becomes more
complicated with deeper shafts and coal-faces at a distance from the pit
bottom, capital investment is required to maintain overall productivity at
about this level. Of course, under particularly favourable conditions with
thick, flat relatively fault-free seams at shallow-depths, much higher
productivities can be achieved, but such cases are balanced out by others

where conditions are difficult and productivities are well below a ton per man-shift overall. Mr J. Langton, in an article in the *Economic History Review,* suggested that the hewers at a Lancashire colliery in the period 1788-99 produced at the rate of 1,357 tons per man-year, but this is improbably high and may indicate a system of work where each hewer has several assistants.[12]

In 1843 there were 12,833 employees on Tyneside, 12,937 on the Wear and Tees and possibly 2,000 elsewhere in Northumberland and Durham giving a total approaching 28,000 for the two counties. This had risen to 33,990 in 1847 and 39,000 in 1852.[13]

Braithwaite Poole estimated the national labour force at 250,000 and the output at 34 million tons in 1850, giving an average of only 136 tons per man-year. His total tonnage figure is almost certainly a gross underestimate, and the manpower an overestimate. Total output was calculated by the Office of Mining Records at 64.7 million tons in 1854 and 92.8 millions in 1864. At the latter date, the labour force was similarly estimated at 307,500 giving an output per man-year of 302 tons. At first glance, this seems rather high, but between 1840 and 1860 some important and widely adopted technical improvements, which will be described hereafter, boosted productivity. Output per man-year averaged 287 tons in the decade 1873 to 1882 but fell back gradually to around 200 tons in the early 1920s.[14]

Accurate figures are available from 1873. Table 1 summarises the position for Great Britain.

Table 1 Manpower in coal mines in Great Britain

Date	No of Workers	
1700	c15,000 – 18,000	
1800	c50,000	
1854	c214,000	
1864	c307,500	
1873-82	503,428	
1883-92	571,719	
1893-1902	732,391	
1903-12	957,848	Annual Averages [15]
1913-22	1,091,391	
1923-32	1,021,692	

The annual averages conceal a sharp decline in the labour force, from 1,248,224 in 1920 to 827,439 in 1932. There was a further gradual decline to just under 704,000 in 1947. The position since then is summarised in Table 2. The reasons for this decline will be considered in a later chapter.

The growth in the labour force in an established mining area prior to the nineteenth century was generally supplied by the natural increase of population of pit villages, and by the employment of an increasing proportion of that population (as when women and young children were

Table 2 Manpower since 1947

Date	No of Workers
1947	703,900
1957	703,800
1960	602,100
1965	491,000
1970	305,100
1973	268,100
1976	247,100

Source: National Coal Board, *Annual Reports*

taken underground). New collieries some distance away from existing ones imported the nucleus of their labour from established collieries, but competed with other occupations for the balance of their requirements.

With the rapid expansion of the industry in inland coalfields from around 1780, particular enterprises could not have looked to their local mining population to supply the whole of the extra labour required. The Earl Fitzwilliam's collieries, as Graham Mee has shown, expanded their labour force from 79 (working four mines) in 1795 to 869 (working seven mines) in 1856. The problem posed was not so great as might at first appear. Taking another example, Thomas North's Cinderhill Colliery sunk in 1841-3, where the original labour force of about 200 was drawn from the older Erewash Valley mines a few miles distant: if, say, twenty mines lost an average of ten men each to Cinderhill, the problem of replacing them would not be great. The natural growth in labour supply in old mining areas therefore very largely made good losses of labour to new mining areas.

There is no doubt that new mines generally paid better wages than old ones, and low-cost housing was also provided. But had recruitment been a great problem, coalmining wages would have risen relative to those of competing occupations to a far greater extent than was the case. Indeed, in the second half of the nineteenth century there was such a relative increase because by then recruitment problems were more acute, and could no longer be met by the employment of females or boys under the age of ten.

The degree of long distance migration of labour prior to about 1850 is impossible to estimate. In the eighteenth and early nineteenth centuries, some Shropshire men and others from the Midlands were enticed by high wages to move to pits in other coalfields where the longwall system was being introduced, and references to Staffordshire immigrants abound in the East Midlands. During the long strike of 1832 in Northumbria, the colliery owners recruited strike-breakers from the metal mines and from other coalfields, Staffordshire again supplying more than other districts.

After 1850, the development of new mining areas where wages were comparatively high and new cottages were provided, attracted men from all over Britain, and a few from Europe. The large new coastal collieries of east Durham, for example, had a 'combination of Lancashire, Cumberland, Yorkshire, Staffordshire, Cornish, Irish, Scottish, Welsh, Northumbrian and Durham accents' in the 1890s as Jack Lawson, MP tells us. He continues:

> The older collieries were more settled in their personnel, but among the great coast collieries there was a constant ebb and flow of the population. A new colliery or a new seam meant bigger money, and there was always an emigration followed by the incoming of new people to take their place. Thus the new collieries were less settled in their personnel, and this fact, together with their large scale operations, produced a different type of people from those of the west, and a different spirit as well.

This applied equally to the concealed coalfield of Yorkshire and Nottinghamshire and to the new mining districts of South Wales.[16]

One of the most notable long-distance migrations was of Lithuanians to the Lanarkshire and Ayrshire coalfield in the early years of the twentieth century. At some mines they formed a substantial proportion of the labour force. They had their own clubs, shops and newspaper and were ministered to by a Lithuanian catholic priest.

WAGES AND HOURS OF WORK

Wages of miners fluctuated according to the state of trade, and varied from place to place. The known examples dating from before the middle of the nineteenth century are too few to be representative, and there is little doubt that the apparent uniformity they suggest is misleading. This broad consideration needs to be remembered in considering the evidence presented below.

It appears that miners in south Durham averaged 5*d* a day in 1461, little if any more than a South Wales example of 1400. In a Durham case of 1530-4, hewers averaged $4\frac{1}{4}d$ a day, but men employed on 'ridding' (cleaning out rubbish) were paid only 3*d*. Miners at Wollaton, Nottinghamshire, were paid 4*d* a day in 1549, rising to 6*d* (and occasionally 7*d*) in 1551-80 when the rate at Coleorton, Leicestershire was the same. At Beaudesert, Staffordshire, 6*d* a day was paid in 1583; and 6*d* to 8*d* was paid at Worsley, Lancashire, in 1600. A group of exploratory sinkers employed by Nottingham Corporation were paid a somewhat exceptional 10*d* a day in 1594-5.[17]

Between the middle of the sixteenth century and the end of the seventeenth, miners' wages approximately doubled, but this did not compensate for the general rise in the cost of living. The Derbyshire justices assessed colliers' wages at 10*d* a day in 1634 and 1*s* 4*d* a day in 1648, while 1*s* a day was paid in Lancashire in 1683. Nef suggests that miners' weekly earnings averaged 4*s* to 6*s* a week in England and 7*s* to 8*s* a week in Scotland in 1699, but the Scottish collier's earnings included those of his bearers.

Duckham similarly estimated Scottish colliers' wages at up to 1s 2d a day, (or even 1s 8d in a few cases) in the late seventeenth and early eighteenth centuries.[18]

That there were wide fluctuations in pay over time is suggested by detailed figures for Leicestershire mines in the early eighteenth century, and that there were considerable variations as between one district and another is suggested by the following examples of hewers' wages: $7\frac{1}{2}d$ a day at Gatherick, Northumberland and 1s 6d at Griff, Warwickshire, both in 1700; and 1s to 1s 2d in Durham in 1708. Similarly, hewers at Barlow, Derbyshire, were paid 1s 6d between 1744 and 1776 and 1s 8d a day in 1780; while hewers at Newbottle, Durham earned about 2s a day in 1788 (when horse-drivers were only paid 1s) and at Measham, Leicestershire, wages in 1791 were between 2s and 3s 6d (plus beer) according to the type of work performed. The Butterley Company's miners were paid 2s 4d to 2s 6d a day in 1795. Even in the same district, wages sometimes varied quite widely. For example, in 1831 when Lord Fitzwilliam's colliers were earning 25s to 26s a week, others in South Yorkshire were said to be averaging 16s. Despite fluctuations it seems clear, however, that real wages rose fairly substantially in the eighteenth century.[19]

Between the 1790s and 1840s, on the other hand, the increase in money wages in the inland coalfields was barely sufficient to maintain the level of real wages. This was a period of fluctuating prices consequent largely on the effects of war. According to Gayer, Rostow and Schwartz's index, (which takes the monthly weighted average of prices of 80 key goods in 1821-5 as 100) the cost-of-living index rose from 89 in 1790 to 154 in 1814 with wide fluctuations between the two dates. The end of the French wars saw a fall in the index to 87.9 in 1822, then a rise to 113 in 1825; it then fell steadily to 84.5 in 1835 rising to 104.3 in 1839, following which it then fell consistently to 73.5 in 1850.

Wages are unlikely to have fluctuated anything like so rapidly as the cost of consumables, so the value of wages in terms of goods must have varied considerably from year to year. Further, the amount of time worked which governed weekly earnings varied with the state of trade. In 1841-2, for example, almost all collieries were on short time for much of the year. Taking the Erewash Valley as an example, holers at Trowell in 1805-6 earned 3s 6d plus an allowance of beer, while in 1829 Alfreton miners earned 2s 6d to 4s 6d a day. Again an agreement at Codnor Park in 1803 gave tonnage prices of 3s 6d for coal and 1s 9d for slack. At these rates a collier would have had no difficulty in earning at least 3s or 4s a day. This was still the average for Butterley and Codnor Park collieries in 1829.

In 1841 the general wage level in the valley was around 3s 6d falling to 3s or so in 1842 and rising to 3s 6d to 4s between 1844 and 1850. Boys' wages in the Erewash Valley undoubtedly fell in real terms, and in some cases in money terms, in this period. For example, John Beasley, a 45-year-old miner at

Shipley, Derbyshire, said that his three sons aged 12, 14 and 16 respectively, and paid 1s 3d to 1s 6d a day were treated much the same as he was as a boy, but were not so well paid. Similarly, William Wardle of Eastwood had been paid 1s a day as a nine-year-old in 1810, whereas his son was paid only 8d a day at the same age in 1841. In 1841 again, hewers in the Leicestershire and South Derbyshire coalfield averaged 3s at Snibston and Whitwick, 3s 4d at Swadlincote and 3s 8d at Moira. A sub-commissioner for Yorkshire gave 20s a week as the average earnings for colliers in his district, but those working thin seams earned 10 to 20 per cent less. Yorkshire colliery owners claimed in 1844 that wages averaged 3s 7½d per day of eight or nine hours, but one admitted that short time at his pits had reduced weekly earnings to 10s. Pembrokeshire wages averaged only 8s to 10s a week in 1841.[20]

Turning to Durham, Galloway says that hewers earned about 4s for six or seven hours work in 1831-2 while putters earned the same amount for an eleven-hour shift and shifters (daily-paid repair men) were paid 3s for eight hours work. Galloway also estimates average earnings for Northumberland and Durham in 1843-4 at 3s 8d. Greenwell gives the normal rates in these counties in 1849 as 3s 9d to 4s 3d for hewers and 3s for shifters, in both cases for an eight-hour shift; about double the Durham examples of the 1780s quoted above. There would, however, be considerable fluctuations in the intervening period as in the West of Scotland where colliers' wages are said to have advanced from 4s a day in 1799 to 5s a day shortly after, continuing thus until 1815 or 1816 when they began to fall reaching 3s 6d in 1823. In most districts, colliers earned between 3s and 4s a shift in the 1840s.[21]

Until the second half of the nineteenth century, miners generally earned little more than workers in most other unskilled and semi-skilled occupations except when there was an exceptional rise in the demand for coal. Thus, two men employed by Lord Fitzwilliam as garden labourers in 1795 at 1s 2d to 1s 6d a day took jobs as miners and earned 2s in one case and 1s 6d in the other. The wages of another increased from 1s to 1s 6d. In the Erewash Valley, 2s a day was the usual rate for a farm labourer (except at harvest when pay was higher) in the 1830s and 40s. Similarly, the average paid in the West Riding for this class of work in 1851 was 14s a week. However, the higher daily rates paid to coalminers than to agricultural and horticultural labourers were partly offset by the greater regularity of employment in agriculture.[22]

Some occupations in mining areas, like frame-work knitting in the East Midlands and hand-loom weaving in Yorkshire, Lancashire and elsewhere, were notoriously ill-paid and yet few who worked in these trades were attracted to coalmining until after mid-century. Despite their low wages, they valued their independence. Again, many were no doubt afraid to ride the shaft before the cage was adopted. Colliery proprietors were not over-anxious to employ undernourished stockingers and weavers in any case, although the demand for mining labour later in the century was such that they had little option. More acceptable to colliery owners were general labourers like those employed on the Thornsett (Derbyshire) Turnpike in 1838 who were only paid 2s a day against the miners' 3s to 4s. In 1796 the Coke family paid the workers at their Pinxton pottery 3s a night for kiln men, and 3s 6d to 4s 2d for the more highly skilled work; as against the 3s to 3s 6d paid to their colliers.[23]

One study which compares mining wages at Griff, Warwickshire in the period 1701-29 with the wages of others, suggests that colliers, at 1s 4d to 1s 6d a day, were paid slightly more than the most highly paid craftsmen (free masons), and twice as much as common farm labourers. However, the figures for non-mining labour are of justices' assessments for the whole of Warwickshire from 1684 to 1729, and not of actual wages paid, which are likely to have been rather higher. Further, wages at Griff Colliery were markedly higher than at most collieries, indicating that there was a good market for its coal. In 1701, it had a labour force of 171 of whom 78 worked underground, and during the succeeding three decades agents were sent to Shropshire to recruit additional labour. Similarly, two Leicestershire mines where wages were also high, Measham and Swannington, recruited men from Shropshire, Derbyshire and Warwickshire (some of them referred to as 'Griffmen') in the same period. Recruits from other coalfields were normally paid more than local men, being given lodging allowances and sometimes binding money.[24]

Miners in 1844 in West Yorkshire averaged 3s 7½d a shift when on full-time (although short-time could reduce weekly earnings to about 10s) while male slubbers in the Leeds woollen district earned £1 9s a week, male dyers earned from 16s to 18s a week and dressers earned £1 0s 8d a week. In 1841, the miner in the Ashby-de-la-Zouch district was clearly better off than a farm labourer. A farm labourer at Measham earning 12s a week (plus 5s a week extra at harvest) had a much more restricted diet than a Moira collier earning 18s a week or more, depending on the state of trade. In Flintshire, fillers earned 15s and cutters 16s a week in 1850 rising to 18s and 20s a week respectively by 1860, while the weekly earnings for brick-makers rose from 12s to 13s for unskilled, and from 18s to 18s 6d for skilled men over the same period. The weekly earnings of farm labourers rose surprisingly from 8s to 14s.

In the second half of the nineteenth century, an ever increasing demand for coal caused miners' wages to improve relative to competing employments. In 1856, colliers employed by one of the largest companies in the Midlands (Butterley) earned up to 5s a day (between 15s and £1 a week) while farm labourers earned 13s or 14s a week, surface labourers about the same, enginemen between 15s and 18s, joiners and carpenters 20s, wheelwrights 21s or 22s and banksmen 18s to 20s. However, by 1900, colliers in the same area were paid 7s 10d or more a day, compared with about a pound a week for farm labourers and 36s a week for skilled craftsmen; and by

1914, underground workers had improved their comparative position even further. Piecework colliers in the Leen Valley then averaged 9s 8¼d a day and daywage colliers 8s 2¼d a day against 5s 6¼d for a skilled craftsman on the surface. The Lancashire cotton, and Yorkshire woollen mills did, however, pay their male employees rates fully competitive with coalmining.

It is not suggested that miners' wages rose continuously between 1850 and 1914. On the contrary they rose rapidly during booms and fell back during slumps. For example, between 1871 and 1873 (the so-called 'coal famine' period) wages rose in most districts by at least 50 per cent, but the whole of this gain was lost between 1875 and 1880. Again, wages rose slightly in 1881-3 with the temporary improvement in trade, and fell back in 1884. Between 1888 and 1890, the miners of the inland coalfields secured increases on their basis rates of 40 per cent. This was reduced to 30 per cent in August 1894, but never subsequently fell below this. In some coalfields (South Wales, Northumberland and Durham in particular) wage rates were tied to the price of coal, and fluctuated frequently in consequence.

The cost of living fell very substantially between 1882 and 1895, so the real rise in earnings was greater than the increase in wage rates. Miners in this period paid a small fixed charge for coal, and most also had low fixed rents (indeed, most miners in Northumberland and Durham lived in rent-free cottages). Foodstuffs bulked large in the family budget, and according to the Co-operative Wholesale Society (CWS) an average family grocery order, which would have cost 7s 6½d in 1882, cost only 5s 11d in 1888, and 4s 10½d in 1895. From this point, the CWS index rose, the cost of the average grocery order being 5s 4d in 1900 and 6s 4½d in 1914. However, miners' wages more than kept pace with this rise although they continued to fluctuate with the state of trade. For example, the average wages per shift of piecework coal-getters in Northumberland rose from 5s in 1879 to 8s 5d in 1914, while in Nottinghamshire and Derbyshire they rose from 5s 4d in 1888 to 9s 10d in 1914.

At Govan Colliery (Scotland) colliers' weekly earnings rose from 14s 8d in 1879 to £2 15s 2d in 1913. It has been computed that, for Britain as a whole, the wages of piecework coal-getters were 86 per cent higher in 1914 than in 1888, the wages of firemen (ie supervisors) were 70 per cent higher; the wages of putters (haulage hands) were 78 per cent higher and those of underground labourers were 88 per cent higher. While groceries cost slightly more in 1914 than in 1888 (according to the CWS index) fuel and rent for most miners cost the same, so the increase in real wages was of the same order of magnitude as the increase in money wages over the period as a whole, though there were considerable seasonal and cyclical variations within the period.[25]

Taking the whole period covered by this short survey, the standard of living of miners measured in material things improved enormously, and Table 3 gives some indication of the trend. The cost-of-living index used here, compiled by E. H. Phelps-Brown and Miss S. V. Hopkins, is based on only a narrow range of goods. Indexation always distorts, as can be demonstrated by taking an alternative base period, but other factors causing distortion here are the narrow range of goods included in the index, the inadequacy of the wages data, the failure to take account of time worked and supplementary income and seasonal and cyclical fluctuations in wage rates and prices. Nevertheless, the table has some use in indicating the broad long-run trend.

Table 3 Typical day's wage of a skilled collier in the East Midlands

Year	Day Rate (Sterling)		Cost-of-Living Index (1451-75 = 100)*	'Real' Wage Index (1914 = 100)
	s	d		
c1550		6	200	29.3
c1650	1	0	700	16.7
c1750	1	6	600	29.3
1790	2	6	871	33.6
1805	4	0	1521	30.8
1830	3	6	1146	35.7
1840	3	6	1286	31.9
1842	3	0	1161	30.2
mid–1844	4	0	1029	45.5
1848	3	6	1105	37.1
1856	4	6	1264	41.7
1888	5	4	950	65.3
1914	9	10	1147	100.0

*Based on Phelps-Brown, E.H. and Hopkins, S.V., 'Seven Centuries of the Prices of Consumables, Compared with Builders' Wage-Rates', *Economica*, XXIII, No 92, 1956, cited in Burnett, p199.

The rapid expansion of coalmining after 1850 coincided with the equally rapid run-down of knitting and hand-loom weaving, and in the last quarter of the century agriculture was also depressed, so mining gained much labour from these occupational groups. Comparatively high wages were not the only attraction, however. Colliery owners also provided low-cost housing for many of their employees, and home coal was either free or very cheap.

In the nineteenth century, few mining employers paid men who were away from work through sickness, although the better employers did pay 'smart money' to men injured at work. The Children's Employment Sub-Commissioners found widely varying practices in 1841-2. For example, the Butterley Company (Nottinghamshire and Derbyshire) had a sick club to which all employees earning 8s or more a week were required to contribute 1s a month. When they were off work through illness or injury, they were paid 6s a week. The company made good any deficiency in the fund, and also paid for medical assistance in cases of serious injuries at work. On the other hand, at Thomas North's collieries nearby, a fund to pay for periods of absence through injury had been started only a few weeks previously by the men themselves. The rate of contribution was only 2*d* a week for men and 1*d* for boys. One of North's employees, a 16-year-old boy, had been buried by a fall of bind and had suffered twelve weeks absence from work without a penny from his employer. Paternalistic employers, like the Lords Middleton in Nottinghamshire and Fitzwilliam in Yorkshire, did pay 'smart money' (injury pay) and indeed Fitzwilliam was reported to be the only employer in the Rotherham area in any industry who did so. In Northumberland and Durham 'smart money' was said to vary between 4*d* and 2*d* a day. An agreement of 20 May 1876 fixed 5*s* a week as the general level of 'smart money' in Northumberland (and presumably Durham) and this amount was still current in 1900. Surprisingly, South Staffordshire was another area where 'smart money' was paid (6*s* a week from the employer and 6*s* from the men's fund in 1843). The most likely reason for this is that the risk of an accident was so great in this district that it was impossible to recruit labour except by paying 'smart money'.

Many miners contributed to friendly societies, both the well-established 'orders' and also small local societies, based, very often, on public houses. Colliery sick clubs or 'field clubs' became increasingly general from about the middle of the nineteenth century, although one at least traced its establishment back to the eighteenth. This was at the Workington and Harrington collieries, where, since 1783, there had been a compulsory deduction from wages of 6*d* a week to which the employer added a third. Benefits were paid for absence through sickness or accident and also on the death of the contributor or a member of his immediate family.

By 1870, widespread criticism of 'field clubs' was made by the men, because they had no say in running them. Gosden believes that 'pressure exerted by miners' unions and the Truck Act of 1861 helped to put an end to the pit clubs and clear the way for the development of Miners' Permanent Relief Societies', of which the first was founded in Northumberland and Durham in 1862. In fact, many 'field clubs' continued, but under the control of workmen's committees, and indeed some even exist today.

Of course, an injured workman who could prove negligence on the part of his employer might be able to recover damages at common law. In practice, there were various defences open to an employer and few workmen could afford legal advice anyway. The Employer's Liability Act of 1880 had only limited usefulness, but the Workmen's Compensation Acts of 1897 and later, made provision for compensation payments for personal injury by accident arising out of, and in the course of, the employment, whether or not there had been any negligence on the part of the employer.[26]

Before the Industrial Revolution, twelve hours seems to have been the normal length of the working day. It was so at Wollaton in the early seventeenth century, and at Griff in the early eighteenth. Indeed, even in

1842 this was still the most common shift length, taking the country as a whole, but in the Erewash Valley thirteen to sixteen hours were common. Long shifts were also worked around Oldham. In Durham and Northumberland, boys worked much longer hours than the colliers, in some cases thirteen or fourteen hours. Usually, one shift of boys had to service two shifts of colliers. In 1849, when eight hours was regarded as the normal shift for a man, 'trappers' (young boys employed to close ventilation doors) worked twelve.

There seems little doubt that the lengthening of the working day, where it took place, was a response to the increasing demand for coal, and indeed evidence for the Erewash Valley in 1842 shows this lengthening of the day in process.[27] Similarly, the employment of women and young children underground in Scotland and some English districts, was necessitated by an expansion in demand. The growth in the size of the individual coalmine in the eighteenth century necessitated improved ventilation, and this provided work (closing doors) which young children could do. 'Trappers' were often only six years old. Again, the introduction of asses and ponies on underground roadways in the second half of the century provided work for slightly older children, and reduced the need for adult barrowmen. The employment of women and children to some extent offset labour shortages which developed during the century, becoming acute from about 1780.

Fig 2   Putter or haulage hand in the Halifax district, 1842 *(I. K. Griffin)*

Women and girls were employed underground in Scotland in considerable numbers in the seventeenth century, mainly as 'bearers' carrying coal along the roadways and up the pits in baskets. The employment of females underground in England was, however, exceptional before the eighteenth century. Even then, the numbers employed in Durham, Northumberland and Cumberland, where the demand for labour was greatest, were small, and the practice had died out in these counties by about 1780. In 1842 females were employed below ground only in Scotland,

Therefore be ye also ready.
Matthew XXIV Chap. 44 Verse

The mortal remains of the Females are
deposited in the Graves at the feet of the
Males as undernamed.
1st Grave beginning at the South end.
Catharine Garnett Aged 11 Years.
Hannah Webster Aged 13 Years.
Elizabeth Carr Aged 13 Years.
Ann Moss Aged 9 Years.
2nd Grave.
Elizabeth Hollings Aged 15 Years.
Ellen Parker Aged 15 Years.
Hannah Taylor Aged 17 Years.
3rd Grave.
Mary Sellors Aged 10 Years.
Elizabeth Clarkson Aged 11 Years.
She lies at the feet of her Brother James Clarkson.
Sarah Newton Aged 8 Years.
Sarah Jukes Aged 10 Years.

Plate 6 Memorial at Silkstone, Yorkshire, to 26 children drowned by an inrush of flood water, 4 July 1838. One of the girls was only 8 years old. *(E. A. Dyson)*

parts of Lancashire and Cheshire, the West Riding of Yorkshire, and South Wales. In most other areas it had never been the practice to employ females underground, though odd cases may be found. (For example, in 1789 Shirtcliffe Bros employed ten men, four women and two girls underground and one man and six women above ground at Nethermoor Lane, Eckington, Derbyshire, which is near the Yorkshire border). The Coal Mines Act of 1842 made it illegal to employ females or boys aged under 10 below ground after 1 March 1843, but this law was widely evaded for some years. As an example, Aaron Stewart who was a founder member of the Nottinghamshire Miners' Association began work at Coleorton, Leicestershire, in 1853 when he was only eight years old. Scottish miners particularly resisted the implementation of the law because it robbed them of the unpaid services of their wives and daughters and considerably reduced their family earnings.[28] The manager of Redding Colliery posted a notice informing his miners that no females were to go underground from 1 March 1843, but he found it necessary to repeat the injunction in 1845.

NOTICE.

NO FEMALES

Permitted, on any account, to work under ground at this Colliery; and all such is STRICTLY PROHIBITED, by Orders from His Grace the Duke of Hamilton.

JOHN JOHNSTON, Overseer.

REDDING COLLIERY, 4th March, 1845.   J. Duncan, Printer, Falkirk.

Fig 3   Notice banning females from working underground at Redding Colliery, 1845

'TRUCK' AND DEDUCTIONS FROM WAGES

In the nineteenth century and earlier, colliers on piecework often claimed that they were not credited with all the coal they had produced, while owners alleged that the men practised deceits on them. When the measure was the corf, the men could arrange lumps of coal as to make a half-full corf appear to be full. They were also accused of sending up slack and poor coal instead of discarding them. When such offences were discovered punitive deductions were made from wages. For example, if more than a certain proportion of slack was included in the corf, the whole might be confiscated. Such a rule was in force at Cinderhill (Nottingham) as late as 1867. In the North East, an insufficiently filled corf was 'set out' (confiscated) and one containing dirt or foul coal was 'laid out' — a fine being levied according to the proportion of the inferior material. The men claimed that it was not possible to distinguish foul coal from good by the light of candles or Davy lamps, so they were punished for what they could not help. Setting out and laying out could be quite arbitrarily and unfairly applied, as the men complained in the 1832 and 1844 strikes. Thus, between 11 July and 13 December 1843, at Ann Pit, East Cramlington Colliery, 6,249 corves of 7cwt capacity were laid out and at this colliery laid out corves were confiscated as well as those set out. Here any corf containing less than $6\frac{1}{4}$cwt was set out.[29]

The 1872 Coal Mines Act provided that where miners were paid according to the amount of mineral produced, the unit of measurement should normally be the imperial ton and not a capacity measure or 'long' ton. An additional safeguard for the men was the right to appoint their own check-weighmen; although even that was not an absolute guarantee that they would be paid for all the coal produced.

Comprehensive fines were levied at various places and periods. Penalties for absenteeism were common: a Nottinghamshire collier of 1848 was to be fined a shilling for each time that he neglected his work 'through drunkenness or idleness'; while at Staveley in 1866 the penalty for a day's absence was a sum not exceeding 5*s* — equivalent to an average day's pay. The Blackwell contract rules of 1877 contained this sweeping provision:

> The Agent or Manager may impose any fine which he thinks reasonable, not exceeding Ten shillings, on any Stallman or Labourer, for any breach of these Contract Rules.

Fines for not conforming to standards laid down by management, or for damaging or misusing colliery equipment, or for some other breach of duty, had been similarly enforced in the seventeenth century.[30]

In the eighteenth century, the payment of a small amount each week (called 'subsistence') with the balance of wages being carried forward and settled periodically either in cash or kind, was a widespread practice. At Griff, weekly 'subsistence' in the early part of the century was usually about 2*s* and tickets encashable or exchangeable for goods, at a future date, were issued to cover the balance. The colliers obtained the bulk of their provisions at their owner's truck shop, where they were charged 'the full market values'. Truck remained common until well into the nineteenth century. Often, high prices were charged for shoddy goods. Also, the owner could adjust real wages, while leaving money wages unchanged, by altering prices in the 'tommy shop'. Further, the system enabled owners to obtain credit both from the suppliers of the goods and also from the men who gave their labour in advance of the payment of wages.

Truck was worst in those districts where the butty system prevailed. Charles Morton, Mines Inspector for the Midlands commented in his 1851 Report that: '"Butties" are often directly or indirectly connected with taverns or shops, where the miners' earnings are spent in the purchase of bad and dear ale and provisions.'[31] Many butties paid wages in these taverns, and insisted on the men buying a large quantity of beer before handing out the balance of wages.

The 'buildas' system may be regarded as a variant of truck. In the nineteenth century, it was in full operation only in Staffordshire, although there were traces of it elsewhere in the Midlands. It was suggested to the Midland Mining Commision of 1843 that the system was derived from an ancient custom of Buildwas Abbey whose tenants were supposed to be required to give labour services for an allowance of beer. It may be significant that the beer allowance was a charge against the mineral owner and not the lessee or contractor who worked the pits. Galloway saw this system as evidence of a former servile status, and he pointed out that in the same district haulage hands underground were called 'bondsmen' or 'bandsmen' and were collectively known as 'the band'. The 'buildas' itself was a requirement that the colliers should work several hours (6am to 11am according to one observer) for a beer allowance; the men had also to work for nothing while 'cleansing' their places of work, and on 'quarter days' they were given only a quarter of a day's pay for half a day or more of work. It was said of the South Staffordshire butties that they supplied bad ale although the owners allowed them $3d$ a quart for it; and in a Nottinghamshire example, the mineral owner had $2\frac{1}{4}$ per cent deducted from the royalty rent due to him to pay for the colliers' drink under the terms of a 99 year lease which expired in 1837, despite the fact that the beer allowance had been stopped some years earlier.[32]

T. S. Ashton regarded truck primarily as a means of overcoming a shortage of coin. This shortage should have been, and sometimes was, met by issuing trade tokens or by paying men in relays. Unlike these two methods, truck provided a subsidiary profit for owner or butty. It is true that some philanthropic employers, like the Lords Fitzwilliam in South Yorkshire, and Moira, Ferrers and Stamford in Leicestershire, sometimes supplied their men with wholesome food below market prices, especially when food prices were high or when trade was slack, but this should not be allowed to conceal the essentially exploitative nature of the truck system. Philanthropy was, in any case, partly an insurance against riots.[33]

The Truck Act of 1831 which made the practice illegal was widely ignored, and this was one of the chief grievances of Midland miners in 1844. The strike of that year probably did more to bring truck to an end than the several subsequent Truck Acts. Truck had virtually disappeared from Northumberland and Durham before 1840, and few cases are heard of in the East Midlands and most other coalfields after 1850. However, tommy shops and payments of wages in public houses lingered on in South Staffordshire (and to a lesser extent the Swadlincote area of South Derbyshire and a few other localities) into the 1860s and 70s. For this, the continuance of the butty system was largely to blame.

MINERS' TRADE UNIONS

Miners were active in many food riots in the eighteenth and early nineteenth centuries (like the Nottingham cheese riots of 1766), although they do not deserve the reputation of 'the most active of the insurgents' given to them by Ashton and Sykes. In the East Midlands at any rate, frame-work knitters were generally far more riotous than miners.[34]

Towards the close of the eighteenth century, strikes of a recognisably modern character, directed at increasing wages or correcting grievances, were taking place in some coalfields. In the West of Scotland widespread strikes were organised in the 1790s by temporary combinations whose members were bound by solemn oaths; a process called 'brothering'. In 1765, the yearly bond was the cause of a long strike by Durham miners, and it was a contributory cause of further strikes in the early nineteenth century.[35]

The first miners' trade union with an air of permanence about it was led by Tommy Hepburn of Hetton and lasted from 1830 to 1832. It was called The Coal Miners' Friendly Society. Most miners in Durham and Northumberland joined it and came out on strike in April 1831. Their demands included the abolition of truck and a shortening of the hours of labour of boys, but the grievance which called the union into being was the withdrawal of the guaranteed minimum wage formerly specified in the annual bonds. It is interesting, in view of the harsh criticism made of Lord Londonderry, that a union pamphleter writing in 1844 credited him with ending the 1831 strike:

> . . . at that period our demands were resisted with as much pertinacity as now, our Union was denounced as illegal and tyrannical and our leaders as demagogues. The strike terminated when the Marquis of Londonderry, judging for himself, and resolving not to be hoodwinked by others, broke from your [ie the coalowners'] union, conceded the principal portion of our claims, and thus compelled the remainder of your body to come to a similar agreement.

Wages were increased by about 30 per cent, the terms of the bond were improved and boys' hours reduced from 14 to 12 a day, but in 1832 the men were locked out and no man was allowed to resume employment unless he renounced the union. The owners objected particularly to a union regulation that before a man could start work at a colliery, he must produce a certificate from his previous lodge or be excluded from the union. It was alleged that at Coxlodge, the men refused to work because new men had been engaged by management without presenting a clearance certificate to the lodge committee. Also, the owners objected to a regulation which prevented union members who were hewers from earning more than 4s a day a tactic modelled on the 'darg' of Scotland. Clearly, these regulations were designed to restrict the output of those who were employed. To keep the pits working the owners recruited men from other coalfields, many of whom had no idea that there was a dispute until they reached Tyneside. Many Northumbrian pitmen were evicted from their colliery houses to make room for these newcomers and this added to the bitterness with which the dispute was fought.[36]

Although Tommy Hepburn's union collapsed in 1832, many of its members took part in further disputes during the late 1830s. Then in 1841 the first national miners' union, the Miners' Association of Great Britain and Ireland, was formed at Wakefield under the leadership of Martin Jude. Despite some Chartist involvement in this union, the disputes which followed were, with minor exceptions, unaccompanied by the acts of sabotage, physical violence (and even occasional homicide) associated with previous strikes. Further, the men were motivated by economic rather than political grievances.

The Chartist leadership of the 1842 strike in South Staffordshire cannot be doubted, but even here the report of the Midland Mining Commission rejected the assertion of the colliery owners and butties that the strikers were politically motivated. Instead, the causes were shown to be economic: low wages, truck, buildasses (beer payments), and all the other iniquities of the butty system.[37] The Miners' Association itself is said to have been inspired by Feargus O'Connor, the Chartist leader who became a Member of Parliament for Nottingham, and it was on his recommendation that William Prowting Roberts was engaged as the Association's legal adviser.

During 1843, the miners of Durham and Northumberland deliberately restricted production while W. P. Roberts took into court a series of cases connected with the bonding system. Paid agents, eventually numbering 53 were engaged to recruit members and organise branches throughout the country. By March 1844 the Association had 52,927 members and this figure is thought to have doubled during the months which followed.

When their bonds expired in April 1844 the pitmen of Northumberland and Durham went on strike for a 28 per cent increase in the hewing prices, a reduction of the working hours of boys from 12 to 10 a day, a guaranteed wage for colliers of 15s a week when unable to work through causes outside their control (with lesser amounts for other employees) and bonds lasting for six months instead of twelve. As in 1832, many men were evicted from their homes. The strike lasted for about 20 weeks, and then collapsed, largely owing to the owners' success in importing strike breakers from Wales, Cumberland, Ireland, and the lead mines. Miners in Scotland and the inland coalfields, many if not most of whom were themselves members of the Miners' Association, were engaged in their own local disputes. In many cases, they were locked out for joining the Association, their employers refusing to take them back until they signed a document abjuring trade unionism. By 1848, the Miners' Association had lost most of its members, although there was a temporary recrudescence in 1850-1, and the Association did not completely disappear until 1855. The Association was responsible for a number of permanent improvements in the miners' lot, for example the virtual end of the bond system in the North East and a reduction in hours of work in the East Midlands.[38]

Following the collapse of the Association, informal clubs and committees at pit level kept up an agitation on the need for closer state regulation of the industry in safety matters, shorter hours, the welfare of young miners, the abolition of truck, and true weighing. An ex-miner from Scotland, Alexander Macdonald, was a florid propagandist who provided a focus for these local bodies. He helped the local societies to organise, won much public support for the miners and initiated legislation to improve the miner's lot. In 1874, he entered Parliament at the same time as Thomas Burt, leader of the Northumberland miners, and they are usually regarded as the first two 'labour' MPs although they sat as Liberals. In 1858, the South Yorkshire Miners' Association came into existence and was quickly involved in a dispute at the Fitzwilliam collieries where a 5 per cent wage reduction had been ordered. Earl Fitzwilliam closed the pits until the men agreed to leave the union, which they did at the end of eight weeks. The forty or fifty 'ringleaders' were not allowed to resume work. Pit organisations in many other districts were also functioning as trade unions. For example, at Cinderhill Colliery, Nottingham, prior to 1861, the miners' lodge restricted output to 600 tons a day; and in 1862 they struck for higher wages, weekly instead of fortnightly pay, and the right to appoint checkweighmen. They had a well-organised strike fund and stood out for some months.[39]

In 1863, the Miners' National Union or Association (MNU) was formed under Macdonald's leadership. It was a loose federation of district associations. The MNU soon gained members in most coalfields and became engaged in numerous disputes, principally in 1867-8 and 1871-5. Many, if not most, of its leaders were Methodist local preachers, William Brown of Hunslet perhaps being the most colourful. His meetings were like revivalist services. In 1869, a rival, more militant, body called the Amalgamated Association of Miners (AAM) led by Thomas Halliday was set up. Its original strongholds were Lancashire, Staffordshire and South Wales, but at times it had branches in Cumberland, North Wales, Forest of Dean and Somerset. The Derbyshire and Nottinghamshire Association claimed to have affiliated to both bodies. In the Midlands 'free labour societies' were established by colliery owners in 1867-8, the one at Staveley and Clay Cross being the best known. This took the form of a friendly society and its opening meeting was adressed by John Tidd Pratt, Registrar of Friendly Societies. Charles Binns, general manager of the Clay Cross Company, who had given the earlier union encouragement in 1851, and was to do the same in the future, was paradoxically one of the leading advocates of 'free labour societies' in this period. Their purpose was to undermine the trade union movement. Paternalistic employers like Earl Fitzwilliam also reacted with particular vigour to stamp out trade unionism.

The strikes of 1871-3 were spectacularly successful (except in South Wales) because the coal trade was enjoying boom conditions. Not only did the miners win increases in wages of 50 per cent or more, but they also established their right to appoint their own checkweighmen and they succeeded in reducing permanently the hours of work in most districts. The Mines Regulation Act of 1872, for which the MNU was largely responsible, strengthened the legal position of the checkweighman; provided that the weight of mineral produced should normally be expressed in imperial tons so as to facilitate the calculation of wages; provided for the statutory certification of colliery managers; and raised the minimum age at which boys could be employed underground.[40]

The wage increases gained during the 'coal famine' of 1871-3 were lost in the later years of the decade and in the process of resisting reductions many of the county unions virtually disintegrated. Durham, Northumberland and West and South Yorkshire were the only districts which managed to keep their associations intact, and their membership was sadly depleted.

The MNU was based on strong coalfield unions affiliated to a national federation, and this corresponded to the economic state of the industry itself. The AAM on the other hand sought to weld a multiplicity of weak local associations covering small districts into a unified national structure. The failure of this principle of organisation became apparent in South Wales in the five-month long dispute of 1875. The owners gave notice for a reduction of 10 per cent in wages from 31 December 1874 which was rejected in the Rhondda and Aberdare Valleys (37,920 men) but accepted in Merthyr, Dowlais and elsewhere (19,800 men). A further 20,000 to 22,000 men employed by owners outside the Owners' Association continued working at the old rate of wages. The 19,800 men who had accepted the reduction were subsequently locked out in an attempt by the Owners' Association to crush the union. The AAM, whose membership fell from 106,368 in March 1874 to 34,494 in April 1875, resolved that it could not support the men financially (beyond asking for voluntary contributions from its depleted membership) because so many of them had been unfaithful to the Association and because they had 'struck' without the sanction of the Association; but the MNU which was in a much stronger financial position, levied its members on behalf of South Wales recognising, as Thomas Burt MP (Northumberland) said, that there had never been a more important battle fought on behalf of the rights of labour. He thought that every district which valued union should support the men of Wales. The Welsh Owners' Association unlike those in districts covered by the MNU had refused to go to arbitration, insisting that the men should accept whatever reductions were demanded without any independent assessment and it was this to which the MNU leaders so strongly objected. The ultimate return to work, negotiated with the help of Philip Casey (South Yorkshire) saw the concession of the principle of arbitration but the owners insisted on a sliding scale, tying wages to prices, as part of the settlement.

Partly as a result of the help given by the MNU in South Wales, the AAM agreed to the principle of an inclusive national federation, and fused with the MNU in 1875.[41]

The period 1873 to 1896 is sometimes called the 'Great Depression', but for the coal trade the cyclical pattern of depression and recovery continued, and the upswing of the trade cycle which commenced in 1896 culminated in the Boer War boom. The high and low points are indicated in Table 4.

Table 4 Average export prices per ton (FOB) for coal

Low			High		
Year	s	d	Year	s	d
1879	8	8	1873	20	6
1887	8	2	1884	9	2
1896	8	9	1890	12	5
1905	10	6	1900	16	6

Source: *Colliery Year Book and Coal Trades Directory* (1951 edn), p559

During the mild prosperity of the early 1880s, many of the district unions re-formed, were temporarily successful in securing slightly higher wages for their members, but were reduced to impotence with the renewed depression of 1884-7. The boom which then followed was exploited by the unions of the inland coalfields who co-ordinated their demands on the owners. Between 1888 and 1890, they obtained increases of 40 per cent on the basis of wages ruling in 1888, and in 1889 they formalized their relationship with the formation of the Miners' Federation of Great Britain (MFGB). The owners responded by forming a parallel organisation. The negotiations which took place between the two bodies covered all the major English inland coalfields, but excluded the exporting districts Durham, Northumberland, South Wales and Scotland. Collectively, the inland coalfields were subsequently referred to as the 'Federated Area' (or 'Federated District').[42]

As Arnot has shown, Scotland was difficult to organise because of the wide geographical spread of its mining districts, and of substantial differences in conditions and outlook (especially between Eastern and Western areas). Even Keir Hardie's Ayrshire, the best organised of the Scottish districts, had only one-tenth of the total labour-force of 10,000 in membership in 1889. In 1894, there was a major strike in Scotland and the MFGB insisted, as a condition of giving financial assistance, that the individual districts must federate. South Wales, which did not join the MFGB until 1899, also had a loose federal structure largely attributable to the isolation of the mining valleys. Even when all coalfield unions were in the MFGB, the inland districts (who were its orignal members) maintained their separate Federated District structure within the wider federation.

In Northumberland, Durham and South Wales wages were governed by sliding scales which provided for automatic wage increases and decreases as selling prices rose and fell. The MFGB would have nothing to do with sliding scales. Between 1890 and early 1893 there were considerable reductions in selling prices, and wages in the sliding scale districts therefore fell; but wage rates in the Federated Area were largely unaffected.

Then, on 30th June 1893, the Federated Coalowners requested a 'reduction [of] 25 per cent off present rate of wages' consequent on a 35 per cent reduction in prices since 1890, but the MFGB refused to agree to any reduction and the men were then locked out. One union leader, William Bailey (Nottinghamshire), said that average wage costs per ton were 1s 3d higher than in 1888 but that average pit-head prices were still 2s 6d higher, so that colliery owners were still better off than before the 1888 wage increase. Further, the MFGB considered that they had managed to achieve a living wage level which ought to be a first charge on the industry's proceeds. Pits which could not afford to pay a living wage should close.[43]

At the end of September, the MFGB resolved that where the owners of a colliery agreed to pay the pre-stoppage rates, the men could return to work. In the Leen Valley area of Nottinghamshire, where leading proprietors were sympathetic to the union, production was soon resumed on the pre-stoppage conditions. Within a month, 87,258 men in the Federated Area were back at work supporting with stiff levies the 228,485 men still locked out. Coal was selling at near-famine prices so the owners whose pits were back at work made large profits, as did the owners in those parts of the country (eg Durham and Northumberland) not affected by the dispute. The owners whose pits were still idle were therefore under immense pressure to seek a settlement.

On 13 November, the Prime Minister, W. E. Gladstone, invited the two sides to meet under the chairmanship of Lord Roseberry, Foreign Secretary. They did so, and on 17 November it was agreed that there should be a general resumption of work on the pre-stoppage conditions. The parties agreed that a conciliation board with an independent chairman should be established to determine future wages questions. This conciliation board was in operation almost continuously until 1918. From time to time it fixed limits within which percentage additions to locally negotiated basis rates should fluctuate.

This settlement was regarded as a victory by the miners, but this judgement has been questioned by Dr J. E. Williams. Briefly, miners' wages in this period had two main elements: basis rates and percentage additions. The basis rates were theoretically those ruling in 1888, but in practice were negotiated from time to time at colliery or district level. The percentage additions were decided by the Federated Area Conciliation Board. In addition, pieceworkers received extra payments negotiated on the spot to compensate for abnormal conditions of work done outside the terms of their contracts. Dr Williams concluded from a comparison of coal price movements and movements in the percentage wage variations that the conciliation arrangement was unfavourable to the miners. But this argument, which has been accepted uncritically by others (eg J. H. Porter), is vitiated by the fact that the basis rates themselves, and also abnormality and extra work payments, were subject to frequent variation, and that in this period these payments were on a general upward trend.[44]

The period 1888-1920 was, on the whole, a prosperous one for the mining industry. Output in 1920 at 230 million tons, was 35 per cent higher than in 1888. Meantime, overall output per man per year had fallen from 321 tons to 187 tons while the average pit-head price had risen from 5s 1d to 34s 7d. The war distorted the picture to some extent, but even so the figures for 1913 illustrate the trend: overall output per man, 260 tons; pit-head price, 10s 2d. The explanation for the trend was well put by J. W. F. Rowe in 1922:

> Hence demand has not only increased but it has increased sufficiently to outweigh the increased cost of supply. Prices have risen, principally because of the increase in the value of coal compared with other goods and services. Demand has been the predominant factor in the equation of exchange through the period . . . . This predominance of demand explains the apparent paradox of rapidly increasing wages combined with a fall in the productivity of labour, and it has enabled the industry to pass on to the consumer the burden of that decreased productivity.[45]

The years which followed were to present an entirely different picture, as we shall see.

The Federated District Conciliation Board covered all the major English coalfields except Northumberland and Durham. Even after abandonment of formal sliding scales, the Northumberland and Durham unions continued to be far less militant than the MFGB.[46] Wage disputes in these counties were settled by joint committees of owners and unions (with independent chairmen) and they operated a 'county average' system. As Stanley Jevons explained in 1915:

> The 'county average' is a fixed figure ascertained many years ago for each class of workers, being in Durham 4s 2d for all grades of workers and all kinds of collieries, and in Northumberland 4s 9½d for steam coal collieries, and 4s 7½d for soft coal collieries. The standard hewing prices in each colliery and district of a colliery, are then adjusted from time to time so as to bring the average earnings of that colliery, or one of its districts, within 5 per cent of the county average. If the average is more than 5 per cent below, the workmen will apply to the Joint Committee; if it is more than 5 per cent above, the owners will apply, unless they think that a reduction of earnings would mean curtailing the necessary supply of labour.[47]

These were basis rates theoretically equal to those ruling in 1879, and were subject to percentage additions. Northumberland and Durham also disagreed with the MFGB campaign to secure an eight-hour day fixed by statute. Their leaders argued that it was wrong in principle for Parliament to interfere in bargains between master and man; but a more substantial reason for their opposition was that men in both counties worked less than eight hours already (some as little as six hours at the coal face) while boys worked ten hours. To reduce boys' hours to eight would entail employing more of them to haul the coal underground, and it was feared that their earnings would in effect be borne by the men. By 1907, it was clear that the eight hours' Bill would soon be enacted (as it was in 1908) and Durham and Northumberland then joined the MFGB.[48]

The South Wales Miners' Federation had similarly been admitted to the MFGB ten years earlier after agreeing to abandon sliding scales.[49] In 1902, when their sliding scale agreement terminated, it was replaced by a conciliation board similar to the Federated Area board. However, in South Wales, allowance payments ('consideration') were much more important than elsewhere. As Jevons noted:

> In the last quarter of the nineteenth century there were many mines in South
> Wales in which consideration money was undoubtedly too freely paid. In one
> or two cases . . . it became the custom for the management to make up the
> men's earnings to 4s 9d plus percentage whenever they were short, with
> practically no investigation . . . .

Increasing production costs, an export tax and the recession of 1902-4 made it necessary for managements to seek reductions in labour costs, and they were best able to do this by cutting down 'consideration' payments, sometimes in an arbitrary fashion. The reduction in hours to eight a day in 1909 made cost reduction still more necessary because the drop in output per man per shift (OMS) was in some cases as high as 13 per cent. Cutting allowance payments, especially at pits of the Cambrian Combine where the fall in OMS had been particularly steep, was the cause of considerable unrest which came to a head in 1910.

The Cambrian Combine Dispute originated in a failure to agree on a new price list at Ely Pit, but it developed into a demand for an individual minimum wage of 8s a day for colliers, and 5s for labourers. Mr D. A. Thomas (later Lord Rhondda) on behalf of the owners offered a fall-back rate of 6s 9d to colliers unable to earn more than this on piecework owing to abnormal conditions, and this formed part of a settlement accepted by the MFGB on 15 May 1911, but rejected by the men. The MFGB had been paying £3,000 per week into the strike fund for the Cambrian Combine men, of whom 12,000 were still on strike in July, but they then withdrew their support, and the men went back to work shortly after.

However, the agitation continued. In South Wales, power had passed from the old right-wing leaders into the hands of syndicalists like A. J. Cook, George Barker, and Noah Ablett who formed the Unofficial Reform Committee (URC). The URC issued a pamphlet *The Miners' Next Step* in 1911 which advocated one all-inclusive union for all mineworkers; and deliberate restriction of effort coupled with demands for higher minimum wages and shorter hours so as to absorb the whole of the profit margin, with the eventual aim of workers' control of the industry.

Outside South Wales, the direct influence of the syndicalists in the period

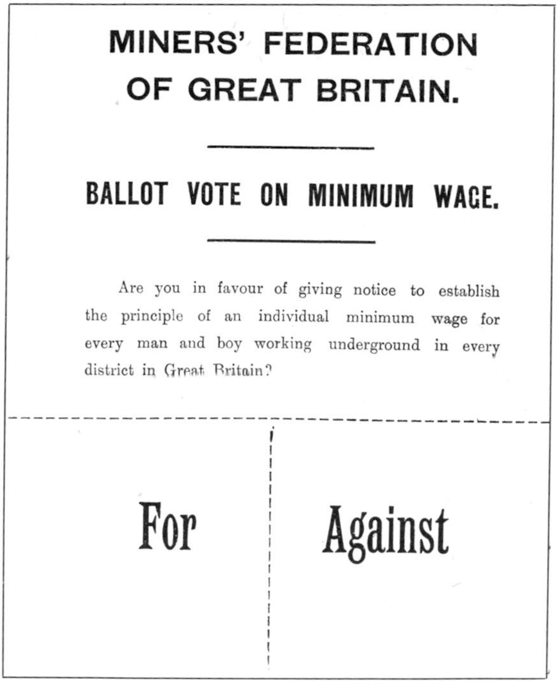

MINERS' FEDERATION
OF GREAT BRITAIN.

BALLOT VOTE ON MINIMUM WAGE.

Are you in favour of giving notice to establish
the principle of an individual minimum wage for
every man and boy working underground in every
district in Great Britain?

For   Against

Fig 4 Strike ballot paper, 1912

before World War I was small. In a period which was marked by widespread unrest, the mining industry outside South Wales was comparatively peaceful. This was no doubt due to the fact that miners continued to improve their comparative wages position.[50] However, one of the demands of the URC triggered off a campaign for a national minimum wage which led to the first national strike of coalminers in 1912, and it is this event more than any other, which is responsible for the miners' undeserved reputation for militancy in this period. The 1912 strike started at the end of February and lasted until early April. As in 1893, the Government intervened, rushing through Parliament a Minimum Wage Act under which statutory minimum wage rates were to be fixed by joint district boards. The miners were disappointed that there was no minimum rate specified in the Act itself, and when a ballot vote was taken 244,011 voted against returning to work, and 201,013 voted in favour. Because there was less than a two-thirds majority for continuing the strike, the MFGB conference ordered a return to work.

The minimum rates fixed by the joint district boards (or their independent chairmen) varied between 4s 9d (in Somerset) and 7s 3d (Nottinghamshire Top Hard mines) for colliers; and between 4s in Somerset and 5s 3d in Lancashire for unskilled underground men. The effect of the Act, as Rowe said:

PINXTON COLLIERIES.

THE MIDLAND COUNTIES COLLIERY OWNERS' ASSOCIATION.

Agreement made between the Midland Counties Colliery Owners' Association and the Nottinghamshire Miners' Association regarding Banksmen's Wages in the Erewash Valley District of Nottinghamshire.

SCALE OF WAGES FOR THOSE WORKING ON THE PIT BANK.

Age	1911 Basis s.	1911 Basis d.	Age	1911 Basis s.	1911 Basis d.
13	1	5¼	18	3	6¾
13¼	1	6½	18¼	3	8½
13½	1	7½	18½	3	9½
13¾	1	9	18¾	3	10¾
14	1	10¼	19	3	11½
14¼	1	11½	19¼	4	1¼
14½	2	0¾	19½	4	2¼
14¾	2	2	19¾	4	3¾
15	2	3¼	20	4	4¾
15¼	2	4¼	20¼	4	6
15½	2	5¼	20½	4	7¼
15¾	2	6½	20¾	4	8½
16	2	8	21	4	9¾
16¼	2	9	21¼	4	11
16½	2	10¼	21½	5	0
16¾	2	11¾	21¾	5	1½
17	3	0¾	22	5	3
17¼	3	2¼			
17½	3	4			
17¾	3	5¾			

1.—The Quarterly advances to be paid at the rate of 1½d. per quarter, and paid to all on the first day in January, April, July, and October.

2.—The Scale to be the Minimum at 1911 New Basis Rates, and any percentage agreed to by the Conciliation Board shall be added to the gross amount worked out on the above Scale.

3.—All able-bodied Men who are now underpaid, and are outside the Scale and have never worked under it shall be raised 2d. per quarter until the standard wage is reached, these advances to be given simultaneously with the Scale advances. (This also applies to strangers).

4.—Pit Top Men and others in responsible positions shall not be bound by above Scale.

5.—Surface Workers over 22 years of age shall be paid current percentage on 1911 rates, and those rates are admitted to be 5/3 per day.

6.—No reduction to be made in the case of any who are now paid above the Scale.

7.—HOUSE COAL ALLOWANCE to workmen being householders, or to a widow, or retired workman with one son or two lodgers working for the Company, 16 Cwts. of Coals for every 20 Shifts made at 3/- per load. Workmen on Compensation to be allowed three loads of Coal, workmen off through sickness, three loads of Coal, until the home coal question is settled for the County.

8.—If the Miners Wages fall below the 1911 Basis this Agreement is open to revision.

9.—This Agreement to apply to all Surface labour engaged on the Pit Banks and Screens.

Signed on behalf of the Owners:
W. B. M. JACKSON.

Signed on behalf of the Workmen:
C. BUNFIELD.

September 12th, 1918.

Fig 5 Banksmen's wage scale in the Erewash Valley, 1918

> . . . varied enormously in different coalfields; South Wales may reckon that
> the struggle was worth while, but many districts, and those some of the more
> important, suffered a six weeks' strike for what was in reality next to nothing.[51]

During the war years, the syndicalists and other left-wing socialists in the
MFGB attracted much support, and by 1918 they were in a dominant
position. In January 1919 the MFGB demanded a substantial wage increase,
a six-hour day and nationalisation of the mines; and in order to avoid a strike,
the government set up a Royal Commission under the chairmanship of Lord
Sankey. This commission recommended a seven-hour day and increased
wages, both of which were accepted by the government and implemented.
On mines' nationalisation, there was no unanimity, but a majority were in
favour of it at the final report stage. However, the government refused to
accept this, but it is not true as has been alleged, that the government
promised to nationalise the mines and subsequently reneged. It was the
interim report which the government promised to honour, not the final
report recommending nationalisation. The MFGB were bitterly
disappointed and tried to induce the TUC to organise a general strike to
procure nationalisation, but the TUC declined to do so.

Coal prices on the export market rose sky-high in the years following the
war and very high profits were enjoyed by the government who were
temporarily in control under war-time legislation. However, when the
export market collapsed in 1921, the government decontrolled the mines
with disastrous results.[52] Faced with the need to reduce costs substantially to
meet the new trading conditions, the owners had no option but to seek wage
cuts, which the men resisted. The lockout which ensued lasted from 1 April
to the beginning of July when the men had to accept substantial cuts in
wages, temporarily cushioned by a government grant of £10 million. Also,
the principle of a national wages pool and nationally negotiated wage rates
which the MFGB had demanded had to be abandoned for the time being,
though it was to be resuscitated as an issue in 1926. Instead, the principle of
basis rates fixed at pit and district level, plus percentage additions, was
retained. Unfortunately for the MFGB, the Federated Area (covering all the
inland coalfields) was replaced by smaller geographical units. In each of
these units the surplus after meeting standard wages, standard profits and
other costs of production was shared between wages and profits in the
proportions of 83 to 17. The workmen's share was expressed as a percentage
on the basis rates and a minimum percentage was specified in the various
agreements. This system (called 'the ascertainment') remained basically
unchanged until after nationalisation.

The miners' unions went into debt in 1921 to keep up lockout pay, and
they had to impose levies to pay off the debts when the men were back at
work. Partly because of this, and partly because they were demoralised by
their defeat, many men became non-unionist. On the other hand, the
extreme left wing achieved dominance in the MFGB and its most famous

member, A. J. Cook, was elected general secretary in 1924, in succession to the 'moderate' Frank Hodges.

A. J. Cook and his syndicalist supporters were opposed by the old 'lib-lab' element in the Midlands of whom Nottinghamshire's G. A. Spencer and F. B. Varley were the most prominent. The syndicalists, who were powerful in a number of unions, sought to use direct industrial action for political ends; while for the 'lib-labs', political questions could only be settled through the ballot box.

A temporary improvement in trading conditions in 1923 enabled the MFGB to secure modest wage increases. However, by 1925 many firms were working at a loss, and a difficult situation was made worse by Britain's return to the Gold Standard at the pre-war parity. As J. M. Keynes pointed out in a brilliant broadside, in the first quarter of 1925 average profits on coal were only about 6d a ton, and to retain her export trade Britain needed to reduce prices by 1s 9d a ton which could only be done in a heavy labour-intensive industry by substantial wage reductions.

The owners proposed not only to cut wages, but also to increase hours from seven to eight a day. Spencer, Varley and Frank Hall of Derbyshire were in favour of some compromise (a reduction in wages, but the retention of the seven-hour day being the most frequently mentioned one) but the overwhelming majority backed A. J. Cook's demand: 'Not a penny off the pay, not a minute on the day'. Recognising that a defeat for the miners would open the way for pay reductions in other industries, the TUC called a so-called General Strike, when the miners' lockout notices expired, on 4 May. However, after only nine days, the predominantly right-wing leadership of the TUC decided to call off the strike because they had no wish to become the *de facto* government of the country. The left, including most miners' leaders, felt that they had been betrayed, and the seven months miners' lockout which followed was conducted in an atmosphere of bitterness.

The miners were once more defeated. Many had drifted back to work long before the dispute officially ended. Some were unable ever to find jobs in mining again. The MFGB suffered breakaways in Nottinghamshire, Leicestershire and Scotland, and a general organisational weakness, while the men had no stomach for further fights. Consequently, in the ten years which followed 1926, the MFGB was unable to defend its members adequately. The owners refused any negotiating rights to the national body. Instead, all negotiations were conducted at pit or district level.

With the increased demand for coal from 1935, and more especially during the war years, the MFGB were able to win back the power they had lost in the 1920s. National negotiations once more took place , and in 1942 the National Conciliation Scheme, which still functions, was established. This was followed, on 1 January 1945 by the establishment of the National Union of Mineworkers (NUM) in place of the MFGB.[53]

4

Technology

AT THE COAL-FACE

In stall-and-pillar work, coal is won by driving roadways ('stalls') through a coal seam. There are several variants of the system, but commonly there are two sets of roadways driven at right angles to each other, with substantial pillars of coal being left to support the roof. In shallow seams, where strong support is needed, this system is still practised and most small private mines today use stall-and-pillar.[1]

The other main coal-winning system, longwall, originated in Shropshire in the seventeenth century. Instead of one man hewing coal on his own in a

Plate 7 Stall and pillar work exposed on an opencast site at Carlcotes, Yorkshire. *(E. A. Dyson)*

Plate 8 Hand holing at Brinsley, Notts, about 1910. *(Rev. F. W. Cobb)*

narrow stall, a number of men work together along a line of face, removing the whole of the coal. In hand-got days, their first step was to 'hole', ie to cut along the floor of the seam with a pick, thus allowing the mass of coal to bear down. Until the late eighteenth century, the coal was won by picks, and wooden wedges driven into the coal face by wooden mauls. The shovels were also usually made of wood. Not until about 1780 were iron tools cheap enough to replace wooden ones.[2]

The roof was supported by a row of wooden props (or 'puncheons') sometimes having wedge-shaped lids to make them tight. As the coal face advanced, the props were removed and re-set in line with it. The waste area from which the coal had been won (usually known as the 'gob' or 'goaf') was supported by props and pack walls built with stones from the roof. Coal was drawn along these 'gob' roads, in baskets ('corves') or boxes, often mounted on sledges, to the main road. There were, indeed, some small mines with low roadways where the sledges were drawn by children on hands and knees to the pit-bottom; and as late as 1843 there were still some Scottish pits where the coal was borne on the shoulders of women in creels not only along the roadways but also up ladders to the surface. This 'bearer' system had been used in some English coalfields, too, in earlier centuries.[3]

The use of wooden rails underground originated in the drift mines of Shropshire in the early seventeenth century; elsewhere, they were not at all common until about 1760 and even then they were usually used only on the main roads, where the corves were trans-shipped from sledges to trams with flanged wheels. Sometimes, wooden boards were laid along the roadways to facilitate the passage of corves, and at Swannington (Leicestershire) in the 1720s coal was conveyed in 'skips' drawn by men along wooden 'rods'; a system intermediate between crude sledging and underground railroads and very similar to the Staffordshire practice. In the large collieries of

Northumberland and Durham, the tram holding one corf gave way to the rolley holding several, at about the end of the century. The Northumbrian corves were large and a crane had to be used to lift them from sledge to rolley.[4]

Flanged iron rails ('tram plates') were invented by John Curr, viewer of Sheffield Colliery, in 1787, and they gradually supplanted wooden rails in underground workings. Joseph Butler, ironmaster of Wingerworth (Derbyshire) was probably the first to use tram plates above ground, which he did about 1788.[5] John Curr also invented shaft guide rails in 1788, but these were little used outside the Sheffield district until the 1840s. Generally, coal was wound in corves swinging freely on a loose rope.[6] In Staffordshire and parts of neighbouring districts (eg Warwickshire and South Derbyshire) where large coal was produced, it was transported in skips. These were wheeled platforms on which the coal was stacked and held in place by iron hoops.

In Staffordshire, too, the Thick Coal, which was particularly subject to spontaneous combustion, was worked by a local variant of stall and pillar known as 'square work'. Here, as Galloway explains, the area to be worked was divided

> . . . into a number of large chambers, or compartments, termed 'sides of work', surrounded on all sides by barriers of solid coal, known as fire ribs, through which no openings are made save such only as are essentially necessary for the extraction of the coal and the ventilation of the working-places. In these chambers the working of the coal is prosecuted on the post-and-stall system, pillars being left to support the roof, and as soon as all the coal that can be got has been obtained, the whole excavation is dammed off by the insertion of airtight stoppings in the few openings made through the barrier.[7]

There were some systems intermediate between stall-and-pillar and longwall. The most important of these was the bank work of Yorkshire,

Plate 9 A steep coalface at Kingshill Pit, Yorkshire, in 1928. Note the use of carbide (acetylene) lamps. (*E. A. Dyson*)

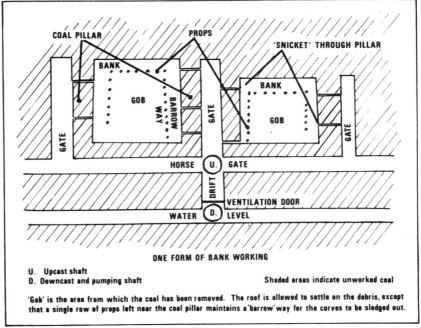

Fig 6   Plan of bank work, pre-1840

and the East Midlands. Here, substantial pillars were left between the coal-faces ('banks' or 'benks') and the roadways. The coal was worked in the same way as longwall, but roadways were not maintained through the gob in the ordinary way. Instead, as the bank advanced, the prop nearest the pillar was left in place and the rest removed; and in this way a small roadway (called the 'barrow gate') was formed next to the pillar. 'Snickets' were driven at intervals between the bank and the main road. The coal was sledged along the length of the bank, and down the barrow way to the 'snicket'. There were several variants on this pattern, one of which is illustrated. In Derbyshire, Nottinghamshire, and probably Yorkshire, true longwall was not introduced until the 1840s.

Bank work, as practised in Derbyshire, entailed a much greater division of labour than other systems. First, the 'holers' undercut the face, inserting small strutts of wood into the cut to prevent the coal breaking away too quickly; then the 'hammer-men' or 'drivers' forced down the coal by hammering wedges into it near the roof; they were followed by the 'remblers' who broke the coal into manageable lumps with hammer-picks; next, the 'loaders' filled the coal into corves and finally, the 'timberers' came along to set props and lids to support the roof. In true longwall, holers were usually a separate class, and in some cases loaders were also recognised as a separate class, although this did not prevent the 'colliers' from helping with the loading. In stall-and-pillar and its variants there was, contrary to the opinion expressed by some historians (eg Professor A. J. Taylor), little division of labour at the coal-face. The pitman working singly in his bord developed an all-round craftsmanship which still, to some extent, distinguishes the Northumbrian miner.[8]

Horses were rarely used underground until the middle of the eighteenth century, when pits in Northumberland and Durham employed them to haul corves on sledges singly to the pit bottom. As we have seen, this gave way shortly after to haulage employing rails and rolleys on the main roads although sledging by hand or pony from bord to rolley way was still the mode. In many East Midlands mines, the roadways (which were usually the same height as the seam) were too low for ponies, so donkeys were used instead.

LIMITING FACTORS

The steep increase in the demand for coal in Tudor and Stuart England, and more particularly in the later period 1750 to 1850, stimulated attempts to overcome major technological problems. In the earlier period, there was a continuous need to find additional coal at shallow depths; and a need to develop techniques for working deeper coal. Tudor miners knew little about geology; but a considerable mystique attached to the 'searchers after coals'. Much of their information was derived from coal outcrops in river beds and theirs tended to be a hereditary craft, so that, for example, several generations of Burtons did the 'searching' in the Coleorton area of Leicestershire. Boring rods, introduced by Huntingdon Beaumont in the

Plate 10   Testing the roof for cracks at Clifton Colliery, Nottinghamshire, in 1895.
(NCB)

early years of the seventeenth century, facilitated the discovery of additional shallow reserves. It was much cheaper to bore a hole than to sink an abortive shaft.[9]

The chief obstacle to working coal at greater depths was water. For seams above the level of free drainage, it was possible to drive tunnels (adits, soughs or water levels) from the workings to the surface, and some quite long ones were driven in this period. A Lancashire example, the Great Haigh Sough driven by Sir Roger Bradshaigh, took seventeen years to complete (1653-70). It was 1,121 yards long, 6 feet high and 4 feet wide, and was ventilated by ten shafts. In the eighteenth century, soughs several miles in length were driven in various districts. Besides draining the workings, soughs also provided ventilation. The improved pumping engines driven by horse, windmill or waterwheel, which Huntingdon Beaumont also introduced, were capable of raising water from below the level of free drainage, but their capacity was limited. In the mid-eighteenth century, James Brindley constructed a waterwheel drainage system at Wet Earth Colliery Lancashire which was the first of a new generation of water-driven pumps. In Scotland, waterwheels were used from about the same time to drive beam engines ('bob-chain pumps') which were quite efficient at shallow depths. But wheels could only be used where there was a good head of water.

After the introduction of the Newcomen engine in the early eighteenth century, drainage was no longer so much a technical problem as an economic one. For small mine owners, long soughs and Newcomen engines ('fire engines') were impossibly expensive, so, in every coalfield, the small owner tried to take advantage of his larger neighbour's drainage works and this gave rise to much litigation and unpleasantness. One reason for the popularity of the bob engine in Scotland was that it was cheap; about £250 in 1790, compared with £1,000 for a small Newcomen pumping engine.[10]

The Newcomen engine did nothing to increase productivity. It enabled colliery owners to work deep seams which could not otherwise have been worked, but with increasing depth more labour is absorbed for a given output. Also, the capital cost of deep sinking can only be justified if the faces are driven much further away from the pit-bottom than was customary with shallow mines, Now, the further the faces from the pit-bottom, the greater is the proportion of 'oncost' to directly productive labour.

More extensive workings necessitated improved ventilation. The late eighteenth century saw the development of new ways of directing the air through the workings, and furnace ventilation. These provided the air for the men and animals to breathe, and diluted harmful gases, but even so, explosions of firedamp were common and many lives were lost in the deep mines of Northumberland, Durham and Cumberland. The safety lamp, invented in 1815, reduced the risk of explosion but did little to save life, because Northern colliery owners then exploited areas of coal which would have been considered too dangerous to work with a naked flame.[11]

Plate 11 A whim-gin at Robin Hood Pit, Rothwell, Yorkshire, about 1935. The building to the rear housed a beam pumping engine. *(E. A. Dyson)*

To say that improved drainage and ventilation did nothing to improve productivity is, however, to miss a vital point. Without them, it would have been necessary to search more thoroughly (and expensively) for deposits near the surface and to work increasingly unrewarding shallow coal. This process could have supplied Britain's needs for coal in the late eighteenth century, but would have absorbed a steeply increasing quantity of labour and land per unit of output. Put another way, there would have been a catastrophic drop in productivity in the absence of the improved equipment and techniques we have mentioned; here we see economies of scale counterbalancing diminishing returns to effort. In considering mining history in any period prior to the twentieth century (with the exception of the period 1840-60 dealt with below) one needs to ask of a technical development not 'to what extent did it improve productivity?' but rather 'to what extent did it help to maintain productivity at pre-existing levels?'

Winding was also a limiting factor partly overcome in the eighteenth century. The earliest winding device, the hand windlass, was the model for horse-gins (cog-and-rung and whim-gins) introduced in the seventeenth century. Improved whim-gins with gearing, and driven by four or more horses, increased shaft capacity. At Walker Colliery about 1750, an eight-horse gin raised a six-hundredweight corf from a depth of 200 yards in two minutes, giving a capacity of 108 tons for a twelve hour shift.

Waterwheels were also used for winding. Huntingdon Beaumont distrusted them because of the uncertain flow of water, but with the introduction of steam power, a ready supply of water could be assured. Both Newcomen- and Savery-type engines were used to pump water for waterwheel winders; and in 1797 there were, according to Curr, thirty or

105

Plate 12 Horizontal double-compound, winding-engine at Elliot Colliery, South Wales, built by Thornewill and Wareham, 1891. *(NCB)*

forty such installations in Northumberland and Durham. By this date, both Newcomen and Watt winding engines were coming into use and they soon displaced earlier systems at all but the smallest shafts. Newcomen-type engines, being easier to maintain, were for long more acceptable to colliery owners than Watt engines. The early winding engines were little more powerful in many cases than the whim-gins they replaced, but more powerful engines came into use after 1840.[12]

COALMINING'S 'INDUSTRIAL REVOLUTION'

The increased demand for coal in the period 1760-1830 saw no dramatic changes in technology such as those which transformed textile manufacture. True, increased use of the steam engine, improved ventilation and the safety lamp facilitated a greatly increased output unaccompanied by the steep fall in productivity which would have been inevitable without them. But the techniques employed in getting, hauling and winding the coal remained substantially unchanged and labour productivity fell steadily. Increasing demand was met by employing more people at more pits for longer hours.

But by the 1830s demand was expanding at a rate which could only be met if there were substantial technological improvements. Without them costs of production would have risen progressively and steeply; and the further expansion of British manufacturing industry would have been severely stunted. The most important limiting factor to be overcome was inadequate shaft capacity. In 1835, T. Y. Hall invented the cage to be used in conjunction with guide rails to hold it steady in the shaft. With this system, first introduced at Glebe Pit, Woodside, near Ryton, the coal was loaded into wheeled tubs (or trams) at the coal-face which were drawn by horses, self-acting inclines (invented by Michael Menzies) or, increasingly, steam-driven haulages, on rails to the pit-bottom where they were pushed on to the cage. The cage itself was conducted up the shaft by guide rails, and finally the tubs were pushed off the cage and on to yet more rails at the pit-top. Underground haulage was simpler and less labour-absorbing, winding was faster and safer, and cages could hold several tubs with greater capacities than the corves oscillating on a loose rope which they replaced. Iron wire ropes replaced hemp, and more powerful winding engines were ordered so that the new system could be fully exploited. Between 1835 and 1850, the winding capacity of a large colliery in the North East more than doubled: from 300 to between 600 and 800 tons a day; and there was considerable potential for further improvement. The shaft capacity of many pits in the Midlands quadrupled.[13] For the miners, riding smoothly in a cage was

Plate 13 Pit bottom constructed in 1851 at Brinsley Colliery, Nottinghamshire. *(NCB)*

infinitely preferable to clinging to a rope swinging freely in the shaft, and this was undoubtedly an aid to recruitment.

Changes at the coal-face were not so dramatic but their importance should not be underestimated. From about 1830, gunpowder came into use at the coal face in many pits. This greatly lightened the burden of breaking down the coal although it increased the risk of explosions.[14]

In Yorkshire, Derbyshire, Nottinghamshire, Leicestershire and South Derbyshire there was an important change in the coal-getting system, from bank work to true longwall. In the type of bank work which had been practised in Nottinghamshire, Derbyshire and Leicestershire a pit had two or three banks of up to 250 yards long and it was only possible to bring out the corves one at a time from each bank. This restricted output severely, few pits worked on the system turning more than 30 to 50 tons a day, and the colliers must have spent a considerable proportion of their time in waiting for empty corves. The much shorter multiple banks of Yorkshire allowed transport to flow more smoothly, but they were subject both to firedamp and spontaneous combustion. The superiority of longwall became generally recognised also in Lancashire (where it partly replaced the Lancashire system combining features of stall-and-pillar and longwall retreating), South Wales, Scotland and other coalfields; and coupled with the adoption of explosives lifted productivity quite substantially at the coal face.[15]

Interestingly enough, the one major coalfield where the introduction of longwall was resisted was Northumberland and Durham. The older mining engineers there believed incorrectly that longwall allowed greater accumulations of firedamp in the goaves than bord-and-pillar. On the other hand following a major explosion at Haswell Colliery caused by an accumulation of gas in a bord-and-pillar goaf, T. J. Taylor, one of the younger engineers, introduced longwall at his East Holywell Colliery to reduce the risk of explosions. As Galloway remarks:

> And towards the middle of the century the long wall, which had formerly been decried, began to be commended instead, by men of high standing, its superiority as regards the simplicity of the ventilation, the closeness of the goaf, the absence of waste of coal etc., being pointed out to the Lord's [sic] Committee in 1849 . . . .

T. J. Taylor recognised that, besides being safer, longwall exploited the reserves more fully than bord-and-pillar:

> The board-and-wall method of extraction is not, however, the one adapted to produce the greatest amounts of marketable coal, and is now being superseded by other methods of working.

However, by the middle of the nineteenth century a whole socio-economic system had been built around bord-and-pillar, and this remained the predominant system in the North East. In some cases, the men went on strike rather than change to longwall, but in fairness it should be recognised that the conservatism was not at all on one side. The bord-and-pillar system itself was improved in this period. John Buddle's panel working, where the area to be exploited was divided into panels surrounded by solid coal barriers, was widely adopted in the 1830s and '40s. This system allowed the coal left in the pillars to be removed without causing a 'crush' or 'creep' (convergence of the strata) to spread through the mine. Nicholas Wood, about 1835, improved this system still further. His method allowed the pillars to be won almost immediately after the bords.[16]

With the change from bank work to longwall in Yorkshire, Derbyshire and Nottinghamshire, the scale of operations and the managerial system were transformed. Up to 1840, few pits there had produced more than 50 tons a day. A large colliery in that coalfield was one having several pits with a total output of perhaps 250 or 300 tons a day. Most winding engines were of the Newcomen type with open-top cylinders (called 'whimseys' in the East Midlands) with nominal horse-powers of between 8 and 25. By contrast, in 1849, Thomas North installed a 180hp engine at his Cinderhill Colliery, Nottinghamshire, to replace the original 30hp engine used when the mine opened in 1843. The winding capacity at this pit with twin shafts of 7 feet diameter each holding one single-deck cage, was then 500 to 600 tons a shift; and a new 10 feet diameter shaft sunk in 1854 (with two cages) had a capacity of over 1,000 tons a shift.[17]

The new East Midlands collieries, Snibston (Leicestershire) and Clay Cross (Derbyshire), both sunk by George Stephenson, and Cinderhill sunk for Thomas North by J. T. Woodhouse, worked coal on the true longwall system. Like Northern mines, they had efficient furnace ventilation, massive pumping and winding equipment, safe, well-constructed shafts, with wheeled tubs, cages and guide rails, mechanical haulage with rails above and below ground, coal screening plant and all the other features of a late nineteenth-or early twentieth-century colliery. They were equipped to produce coal at greater depths than had been usual in the district, and which could hardly have been worked at all by the old methods. Within the space of twenty years (1840-60) almost all the old inefficient collieries had been either reconstructed and re-equipped on similar lines or had closed; and other new and efficient collieries had been opened.

Both in Yorkshire and the East Midlands and also in Durham, the deep, concealed coal measures over to the east of those districts were now being exploited. Without the new developments summarised above, this would have been impossible, and the coalmining industry would, in this second half of the nineteenth century, have been in the process of stagnation and decay. The technological developments in the North East and the East Midlands outlined here were mirrored elsewhere, although a few coalfields were predominantly backward, and in every coalfield some small and relatively primitive mines remained.

The small owner, working shallow seams, whose capital resources are

Plate 14 Vertical winding-engine house at Bestwood Colliery, Nottinghamshire. The steam engine here was built in 1873 and is now preserved. *(NCB)*

Plate 15 The vertical steam winding engine at Stafford Colliery, Stoke Trent. Built by Wren and Hopkins in 1875 this engine was dismantled 1972. *(NCB)*

limited, is perfectly well able to compete with large, deep, well-equipped mines. His ability to do so is no test of his efficiency. As we shall argue later, productivity (the output per man employed per unit of time) is but an imperfect measure of efficiency. The real test of efficiency in a particular case is the extent to which the actual performance measures up to the potential performance under those particular physical circumstances.

Owners whose mines worked the Thick Coal of Staffordshire and Warwickshire at shallow depths were able to operate profitably without any great investment of capital. The Mines Inspector for the Midlands, Charles Morton, in his report for 1851 contrasted these mines with the much more efficient ones in the rest of his district. For example, regarding ventilation he said:

> In the counties of Warwick and Stafford I have descended various mines wherein no artificial means are adopted to produce a circulation of air; the little ventilation there is proceeds from natural causes, and in certain states of the weather these cease to operate.

Similarly, he reported that shaft guide rods or conductors were 'rare' in these two counties.[18]

In South Staffordshire, again, the big butty system lingered on into the twentieth century. T. S. Ashton was incorrect in assuming that this system was a concomitant of longwall coal working. In the East Midlands, it existed long before bank work replaced stall-and-pillar while in South Staffordshire where the system was at its most intense, the Thick Coal was generally worked by a variant of stall-and-pillar called square work; although by mid-century, it is true, some pits had gone over to a variant of longwall retreating pioneered by James Foster of Shutt End Iron and Coal Works.[19] In the East Midlands, the introduction of true longwall brought the big butty system to an end. The colliery owner had too much capital at risk in a large mine to trust it to an unlettered butty. Instead, this period 1840 to 1860 saw a revolution in management as well as technology.

COAL-FACE MECHANIZATION

From mid-century, technological development was unspectacular. On each upswing of the trade cycle new collieries, embodying the latest techniques, were sunk and old collieries were reconstructed and re-equipped so as to be able to work deeper, and more difficult seams. There was a continuing tendency to improvement on underground haulage, with ponies replacing donkeys and also taking over work done previously by boys; and with mechanical haulages taking over work on the main roads previously done either by boys or horses. The ratio of 'ponies' to humans remained at about 1 to 20 until after 1925. There were over 70,000 ponies in 1912 and 65,210 in 1924, but from then the numbers fell steadily to 32,524 in 1938 and 15,858 in 1951. By 1970 ponies had been eliminated at all but a handful of collieries.[20]

Plate 16   Pony putting at Brinsley Colliery, Notts, about 1910. *(Rev F. W. Cobb)*

Similarly, ventilation was steadily improved, mechanical fans gradually replacing furnaces.[21] The last ventilation furnace, at Walsall Wood, Staffordshire, did not go out of use until 1950.

Various devices for undercutting the coal seam mechanically were invented, but could not be successfully applied until a source of power more flexible than steam could be found. Compressed air, first used in 1853 at a colliery near Glasgow, provided the power for a new type of coal-cutter invented in 1863 by Thomas Harrison. This machine had a disc with picks set round its circumference which was caused to rotate just over the floor of the seam as the machine was hauled along the face. It acted like a circular saw, taking a cut (a 'kerf') from the base of the face, thus allowing the weight of the coal to bear down. An electrically driven disc machine was made by Clarke and Stephenson in 1893. The disc machine's greatest drawback was the tendency of the disc to jam as coal fell on it. With electric drives, this could 'blow the fuses' for the whole pit.

A second type of coal-cutter with a similar action but having, instead of a disc, a chain laced with picks rotating on a jib, was introduced in 1873. However, because they were not very robust, chain coal-cutters were not popular until the 1920s. Yet a third type, the bar machine, had as a cutting arm a tapered round steel bar armed with cutter picks. None of these machines was as trouble-free as T. J. Byres seems to think.

By 1911, coal-cutting machines were used at 471 British mines out of a total of 3,100 and even as late as 1938 only 56 per cent of the output of collieries in England and Wales was cut by machine. The percentage for Scotland was rather higher at 68 per cent. Contrary to the view of Byres, Scottish colliery owners had more incentive to mechanise, because their face labour costs were comparatively high and their seams comparatively thin. (It will be appreciated that mechanical cutters produce much less slack than hand holing.) It may be said that the more progressive companies working longwall had gone over almost entirely to cutting and conveying coal by machine by the start of World War II. Bord and pillar does not lend itself so readily to this type of mechanization.

Before the introduction of the face conveyor, coal was drawn from a longwall face through gob roads which were fairly costly to maintain. The face conveyor reduced the amount of roadway maintenance and speeded the movement of coal at a greatly reduced labour cost. Several types of conveyor were introduced in the early 1900s, but by 1930 there were only two types in general use. These were the shaker conveyor, where coal was shaken along a set of iron troughs with a motion resembling a 'cake-walk' at a fun-fair; and the moving endless conveyor belt. At the end of the face, the coal was discharged into trams in one of two ways. In a minority of cases, the gate (or roadway) was sunk below the level of the face, allowing the conveyor to discharge directly into the tram, but the more usual arrangement was the mechanical gate-end loader.[22]

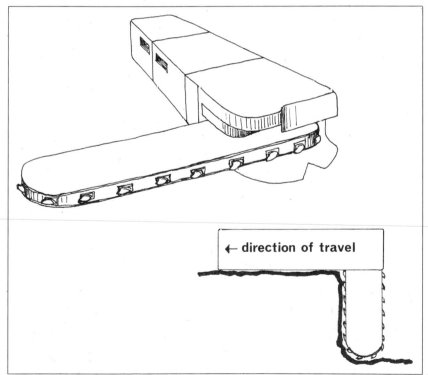

Fig 7   Chain coal-cutter *(I. K. Griffin)*

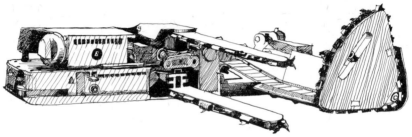

Fig 8   A. B. Meco-Moore cutter-loader (*I. K. Griffin)*

During World War II, a new generation of coal-face machines emerged. The A.B. Meco-Moore cutter-loader, developed by the Anderson Boyes Company from a design which originated in the 1930s, was the first effective power-loader machine. Unlike the earlier prototypes, this machine cut and loaded the coal simultaneously. It had two horizontal cutting jibs like those of a conventional coal-cutting machine, one cutting along the base of the seam and the other a little over half way up; and also a vertically mounted jib (the shear jib) which cut along the back of the web. As the machine moved along, the cut coal was automatically swept up and on to a moving conveyor belt. Before the machine could start its run a 'stable hole' had to be cut out for it by hand at the end of the face. This was a draw-back of all power loader installations until a few years ago.

By 1948, there were some forty or fifty Meco-Moore machines in use. They speeded up the operation and reduced the number of colliers on a face from thirty or more to thirteen or fifteen. However, a Meco-Moore machine was cyclic. It still required a second shift of men to move the conveyor over to the new face line, and to build packs. In competition with the Meco-Moore were various kinds of coal plough, a type of machine pioneered in Germany; and American machines designed primarily for stall-and-pillar work. Neither proved to be particularly suitable in British mining conditions.[23]

Since 1950, there have been two main types of powder-loader machines both of which are non-cyclic, ie they do not require a preparation shift, each new shift starts where the last one left off. The Anderton Shearer is mounted on an armoured flexible conveyor (another innovation from Germany popularly called the 'panzer' or AFC) which is a robust steel structure. The shearer has a horizontally pivoted drum laced with picks around the barrel of the drum. As the machine moves away from the stable hole along the face, this drum rotates at high speed, thus shearing down the coal which is then gathered up and eased on to the AFC by a plough arrangement.

The A.B. Trepanner was originally a floor-mounted machine running alongside the AFC. The trepanning head is like a giant auger cutting into the coal as the machine travels along the face. There are now many variants of both types of machines (eg conveyor-mounted trepanners), and improvements are being made all the time. During the last four or five years considerable progress has been made in eliminating the stable holes. Also machines for ripping the roadways are being used in many places, although the technical obstacles to progress with machine ripping in main gates are proving very difficult.

The longwall retreating system, where roadways are first driven to the boundary of the area to be worked and the coal is then produced from the boundary backwards towards the pit-bottom, is one way of making sure that the advance of the face is not retarded by the ripping operation. This system is becoming increasingly popular, but there are many places where convergence of the strata on the pre-formed gates would be excessive, entailing constant repair work. There is a need to leave wide pillars of coal unworked between retreat faces so as to minimise convergence; but at a time when the capital costs of opening up new reserves are so heavy, the disadvantages of sterilising a considerable proportion of a colliery's take will no doubt impose a limit on the growth of this practice. The supposed advantages of retreat working have, almost certainly, been exaggerated by its advocates.

The development of power loading has been greatly assisted by new ways of supporting the roof. In the nineteenth and early twentieth centuries, wooden props and bars were almost universally used.[24] Rigid iron props and bars were introduced into many mines during the two world wars, though not without some opposition from conservative colliers who thought them unsafe. Since 1945, rigid props and bars have been largely replaced by hydraulic supports. The earliest of these were hydraulic props which operate in much the same way as the hydraulic jacks used in garages. The collier sets the prop vertically and then tightens it to the roof by the operation of a key. The prop will resist pressures of up to 20 tons without yielding. With greater pressures, the prop will yield a fraction of an inch at a time, but without any diminution in its upward thrust.

From these props, hydraulic chocks were developed. A hydraulic chock consists of a number of hydraulic props (called 'legs') having canopies which are steel cantilever beams, the whole being mounted on a steel platform to form one unit. A ram, also operated hydraulically, is attached to the platform. As the power loading machine travels along the face, the AFC and the chocks are moved forward to the new face line by the action of the rams. In some installations each chock is operated separately, but in others they are controlled in batches; and it is possible to control them all from a 'console' of switches in the gate.[25] This improved support system, which in effect provides a steel canopy throughout the face, has facilitated a change from strip packing (building packs of debris in strips along the gob side of the face so that the overlying strata settle gradually) to complete caving, where the gob is allowed to fall in as soon as the face is advanced.

If the period 1840 to 1860 may be regarded as coalmining's first industrial revolution, then the period 1950 to 1970 may be regarded as the second. In

1950, most coal was mined by what was then regarded as the conventional system, with machines used to undercut the face but with the coal got and loaded on to a moving conveyor belt by hand and filled into tubs over a gate-end loader. There were, indeed, some collieries, both longwall and bord-and-pillar, where hand-got methods were used throughout, and this is still true of some small private mines.

By 1970, over 90 per cent of coal was was won by power-loader machines; and output per man-year had risen from 292.8 tons in 1950 (still 40 tons below the figure for 1883) to 456.7 tons. However, one qualification needs to be made: the product has changed very substantially. Before the invention of the steam engine, small coal was unsaleable; thereafter there was a varying sale for it. But at most times, what the colliery owner wanted was large clean coal. Small coal commanded low prices, and often could not be sold at all, so it was often best to leave it underground so as to save the cost of handling it. Even so recently as the 1920s and '30s, colliers in the East Midlands were required to load with a screen (ie a fork) so as to exclude the small coal which had to be cast back into the 'gob'. Power loading machines produce a preponderance of small coal, shearers more so than trepanners. There is, fortunately, a ready sale for this today for power station and industrial

Plate 17   A modern coalface showing shearer power-loading machine, armoured flexible conveyor and self-advancing supports. *(NCB)*

general use. Indeed, much of the product of today's opencast mines is slack and small coal discarded by miners of an earlier age, when up to 50 per cent of the product was thrown into the waste.

MINING DISASTERS AND SAFETY

For the general reader, mining disasters have a fascination which cannot be satisfied adequately here. Our present concern is with those disasters which precipitated improvements in technology.[26]

The most dramatic mine disasters are explosions; and it was as a direct result of an explosion at Brandling Main (or Felling) Colliery, Durham, in 1812 that the Sunderland Society for the Prevention of Accidents in Coalmines was established. In 1815, this society appealed to Sir Humphry Davy to find some safe means of illuminating underground workings; and Sir Humphry immediately set to work. His early experiments demonstrated the low inflammability of firedamp. Firedamp, he said, required a 'considerable heat' for its inflammation, while in burning it produced 'comparatively a small degree of heat'. He therefore concluded that if an explosive mixture be made to pass through narrow apertures of metal the 'cooling powers' of the metal would prevent an explosion. He soon decided that wire gauze was the ideal material since it has many narrow apertures, and does not obscure the flame to an intolerable degree. Writing to the Rev John Hodgson, a leading member of the society and the incumbent of the parish in which Felling Colliery was situated, Davy said on 19 October 1815:

> I have already discovered that explosive mixtures of mine damp will not pass through small apertures or tubes; and that, if a lamp, or lanthorn, be made airtight on the sides and furnished with apertures to admit the air, it will not communicate flame to the outward atmosphere.

This, and similar communications, although intended to be private, were talked about.

That the coalowners were not entirely altruistic in their support for the work of the Sunderland Society is implied by evidence given by John Buddle to the House of Lords Committee in 1829. Asked whether the number of explosions had been diminished by the Davy Lamp, he replied:

> They have, I conceive; but on taking the average for thirty-four years, up to the present period, scarcely one half of which we have had the benefit of this lamp, the loss of life has been nearly about the same; but I attribute that to this cause, that we are working mines, from having the advantage of the safety lamp, that we could not have possibly worked without it . . . .

Without the safety lamp, many mines would have had to be closed once the bords had been worked, so the coal left in the substantial pillars would have been sterilized.

George Stephenson was also experimenting with safety lamps in 1815. His first two lamps were based upon an altogether erroneous theory, and when they were tried in blowers of firedamp they went out. He then designed a third lamp, the order for which was placed on 19 or 20 November, which was based upon the principle already enunciated by Davy. According to Mr J. H. H. Holmes who was present when Stephenson demonstrated his third lamp: 'Mr. Stephenson appeared totally ignorant of the manner in which the air and gases operated upon the light'. Mr Holmes may have been prejudiced but there can be little doubt that the inspiration for Stephenson's third lamp, commonly called the 'Geordie', came from Davy.[27] In this, as in other things Stephenson's genious lay in adapting and improving the inventions of others.

The lamp invented by Sir Humphry Davy had a wire gauze surrounding the flame. If properly made and carefully used, it was safe in an explosive mixture.[28] However, as a mines inspector pointed out in 1851:

> There is good ground for believing that fatal accidents have occured with his lamp through ignorance of its real nature and capabilities.
> Davy recommended a wire gauze, containing not fewer than 784 apertures in a square inch; he advised that the framework and fittings of the lamp should be so arranged as to prevent the possibility of there being a larger external aperture in any part of it; he warned the miner not to expose it to a rapid current of inflammable air unless protected by a shield half encircling the gauze; and he deprecated the practice of continuing to work with it after the wire attained a red heat.
> In my visits to the collieries I have occasionally observed and pointed out to persons present their practical disregard of the foregoing conditions.
> Lamps are now made which contain only 550 or 600 apertures per square inch; the framework is in some instances so loosely put together, or so far dilapidated, as to exhibit even larger openings; shields are seldom met with; and the gauze is seen at times red hot, and smeared outside with grease and coal dust, whereby the risk of ignition is rendered more imminent.[29]

Since Davy's original invention, there have been hundreds of modifications. Some were intended to make the safety lamp more safe; some were to improve the lamp's light output; while others were primarily intended to increase its usefulness as a firedamp detector.

A major innovation was made about 1841-3 by Dr Clanny who had invented the very first effective, albeit impractical, safety lamp in 1811. He placed a glass cylinder around the flame, and a gauze above the glass. In this way, the light output was doubled, although it still gave only half as much light as a candle. This was Clanny's sixth lamp and was similar to an invention of Mueseler, a Belgian, made in 1840. Here, besides the glass cylinder surmounted by a gauze, there was an internal conical metal chimney designed to separate the burnt gases from the incoming air. A Frenchman, J. B. Marsaut, demonstrated that the Mueseler pattern was rather less safe

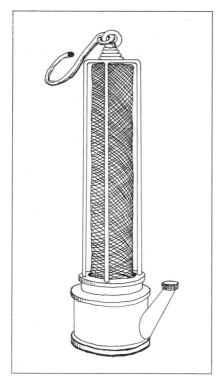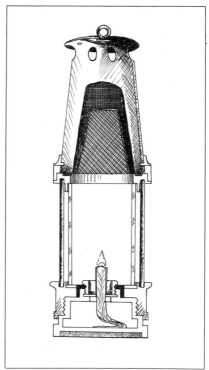

Fig 9 (left) Davy lamp; 10 (right) Marsaut-type safety lamp *(I. K. Griffin)*

than lamps without the internal chimney, and he designed a lamp similar to the Clanny, but with double gauzes. Modern flame safety lamps are based on Marsaut's design.

Glass shields were not adopted by Davy because of the fragility of the glass then manufactured, and it was for this reason that Benjamin Biram used a mica front for his lamp, invented in 1849. While the Biram lamp gave a brighter light than others, it was not favoured because it gave light only in a forward direction, and was of a somewhat flimsy construction. Its cost in 1851 was 7s 6d against 7s for a Davy and 12s for a Clanny.

A Royal Commission of 1886, which examined various designs of safety lamp, specially commended the lamp which came to be known, after modification, as the Ashworth-Hepplewhite-Gray. This had four tubular pillars through which air was drawn to the flame in the brass base of the lamp. The burnt gases passed through a gauze and an external metal chimney. This lamp was recommended for use in strong currents of air. Sometimes, the usefulness of this lamp for gas detection was enhanced by the addition of a hydrogen cylinder. When testing for firedamp, the examiner turned a key allowing hydrogen to pass into the lamp. Firedamp is seen above a flame as a blue cap, and with hydrogen it was particularly easy to estimate the proportion of firedamp in the body of air being examined.

Early flame lamps gave a poorer light than a candle, and in districts like the East Midlands, where there was relatively little firedamp, they did not come into general use until the early years of the twentieth century. Poor light tended not only to reduce productivity, but also resulted in a disease, miners' nystagmus, which caused a painful oscillation of the eyeballs. In a report issued by the Miners' Lamp Committee in 1924, it was recommended that the minimum light output of a flame safety lamp should be increased from 0.3 to 0.8 mean horizontal candlepower. The committee found that there were 403,000 Marsaut-type, 176,000 Mueseler-type and 6,000 other flame safety lamps, and 328,000 electric safety lamps in use in 1923.[30]

Because electric safety lamps, especially those made by Concordia were becoming so popular, the manufacturers of flame lamps introduced important modifications to improve the light outputs of their products. Flame lamps giving 2.5 candlepower and more came on the market in the 1930s, but they were too late to ward off the competition, and they were also unpopular because their casings became very hot. Flame safety lamps continue in use for detecting gas, however, though for some years they have been supplemented first by electric lamps which indicate the presence of firedamp automatically (the chief one being the Ringrose Alarm) and by methanometers, which have now largely replaced the Ringrose.

Since World War II, electric hand lamps have been replaced by cap lamps which direct the light to where it is most needed, while leaving the hands free to work. Roadway lighting has also been vastly improved. Thomas North pioneered the use of gas lamps on the main road of his Cinderhill Colliery in 1861, but apart from their expense, these were too dangerous to be widely used. Electricity was first used for underground lighting at Earnock Colliery, Hamilton (Scotland) and Pleasley Colliery, Derbyshire in 1882; four years later John Davis of Derby installed electric lighting underground at the Mill Close lead mine, Derbyshire. By the end of the century, many collieries were using fixed electric lamps for pit-bottoms and engine houses and subsequently electricity became widely adopted for lighting the main roads.[32]

Following the Haswell Colliery explosion in 1844, where ninety-five employees died, the Government appointed an investigating committee, of which Professor Faraday was the leading member, to ascertain the cause. Faraday concluded that the disaster was triggered off by a minor firedamp ignition, but that coal dust was responsible for the intensity of the explosion and resulting fire. One suggestion made by Faraday's committee, that gas should be drained from the goaves and up the upcast shaft through 12 inch diameter iron pipes, was dismissed as impractical at the time, although it is nowadays standard practice at collieries which produce large quantities of methane.

Despite the work of Faraday and others, the notion that coal-dust causes

Plate 18 A cupola or ventilation-furnace chimney at Flockton, Yorkshire. (*A. R. Griffin*)

explosions was resisted by many authorities. Indeed Dr William Galloway was made to resign from the mines inspectorate after reading a paper to the Royal Society in which he demonstrated conclusively the part played by coal dust in carrying flame along roadways in mining explosions. To their credit, two other mines inspectors, W. N. (later Sir William) Atkinson and J. B. Atkinson, published evidence supporting Galloway's views and in 1893 a Royal Commission accepted the coal-dust theory of explosions. At first, water was used at some collieries to render the coal dust safe, but subsequently the mixing of stone dust with the coal dust became the general practice, and it was required by law shortly after World War I. In recent years, stone dusting the roadways has been supplemented by the erection of stone dust (and, in some cases, water) barriers at intervals along a roadway. Any explosion would release the material in the barrier which would effectively arrest the explosion.[33]

The three main mine gases are 'firedamp', 'blackdamp' or 'chokedamp', and 'afterdamp'. Firedamp explosions have already been considered. Firedamp is a mixture whose chief constituent is methane (CH_4) produced from decaying vegetable matter as the coal seams were being laid down. It is trapped in the molecular interspaces in the coal itself, and in adjoining strata. As coal is worked, the gas is released steadily; but occasionally a considerable quantity of gas (called a 'blower') hisses out of a cavity.

Blackdamp (or chokedamp) is an inert mixture of nitrogen (N_2) and carbon dioxide (CO_2) formed by the oxidation of coal, timber and other materials. Exhaled breath also contributes. It follows that the more

121

blackdamp there is in the general body of air, the less oxygen there is, and it is this which threatens life. Blackdamp is usually heavier than air. In shallow mines, it is a far greater danger than firedamp. Before efficient ventilation systems were introduced, people and horses were frequently overcome by this gas. In the presence of blackdamp the flame of a candle or lamp shortens as the proportion of oxygen in the air falls, and the safe thing to do then is to withdraw. However, witnesses at Erewash Valley pits told the Children's Employment Commission in 1841 that they were not allowed to leave the mine until their candles were put out. Blackdamp (or, to put it another way, shortage of oxygen) reduces efficiency besides affecting health. Those working without good ventilation were always short of breath, a condition known as miners' asthma.

At Hartley Colliery, Durham, in 1862 blackdamp took the lives of 204 men and boys. Here as at many mines in Northumberland and Durham, there was only one shaft divided vertically by a wooden brattice. In the pit bottom, a solid brick wall ensured that the two sections of the shaft were kept separate. A furnace near the pit bottom heated the return air passing from the workings to the upcast section of the shaft. Cold air flowing down the other section of the shaft to make good the partial vacuum was directed

Fig 11   Arrangement of the shaft top with an axial-flow ventilating fan *(I. K. Griffin)*

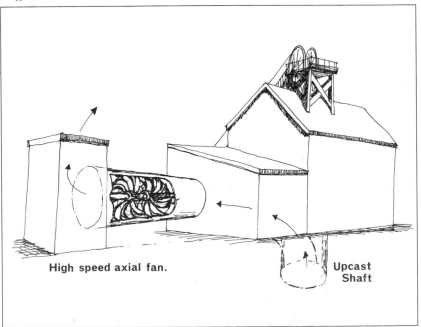

High speed axial fan.   Upcast Shaft

Plate 19 Demolition of a 30ft diameter Guibal ventilation fan built in 1876 at Newstead Colliery, Nottinghamshire, in 1961. (NCB)

Plate 20 A 40ft diameter Waddle Fan at West End Colliery, near Batley, Yorkshire, about 1937. (E. A. Dyson)

through the workings in the usual way by an arrangement of doors and stoppings.

On the day of the accident, the heavy cast-iron beam of the pumping engine at the pit top broke in two, and one half crashed down the shaft, ripping out the brattice as it went and completely blocking the shaft with debris. With the brattice removed, ventilation stopped, and the unfortunate miners gradually suffocated. By the time the rescue teams had cleared the shaft, all the men were dead. This tragedy precipitated an Act of Parliament providing that virtually every mine must have at least two separate means of egress to the surface.[34]

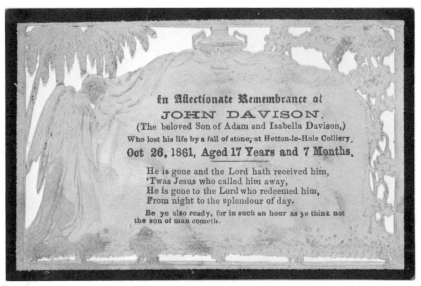

Plate 21   A memoriam card for John Davison, killed by a fall of stone at Metton le Hole Colliery, Durham, in 1861

In the nineteenth century, the term 'chokedamp' was often used colloquially of carbon monoxide (CO) more properly referred to as 'afterdamp'.[35] This is a colourless, odourless and highly poisonous gas resulting from incomplete combustion and is invariably present following an underground explosion or fire. The first report of the Sunderland Society pointed out that many, if not most, of the victims of the explosion died not from the explosion itself but from the afterdamp poisoning. Similarly, Faraday said that nearly all the ninety-five victims of the Haswell disaster were killed by the afterdamp. But it is only within the last decade that miners have been issued with self rescuers. These are gas masks to be worn while withdrawing from the workings through concentrations of carbon monoxide, and act by converting carbon monoxide into carbon dioxide.

A recent disaster, where the eighty victims all died from carbon monoxide poisoning , was at Creswell Colliery, Derbyshire in 1950. This was a fire caused by frictional heating of a torn rubber conveyor belt. Even before the disaster, manufacturers were developing belting with fire resistant properties, and the disaster brought home the urgency of this matter. Consequently PVC soon replaced rubber, and fire-proofing was applied to the cotton duck foundation of the belt.[36]

Similarly, lessons have been drawn from other types of accident such as inrushes of water, falls of roof, and shaft accidents. There has, indeed, been a dramatic improvement in the coalmining industry's safety record. In the twelve years ending 31 December, 1867, there were 12,590 persons killed in colliery accidents in Great Britain. In 1866 alone, there were 1,484 fatalities although 1867, with 1,190 fatalities was markedly better. There were 333,116 coalminers employed in 1867, so the fatal accident rate per 100,000 men employed was 357. The corresponding figures for 1885 are 1,214 fatalities at a rate of 216 per 100,000 men employed. Subsequently, there was a progressive reduction to 460 fatalities at a rate of 62 per 100,000 men employed in 1949. In 1970-1, there were 92 fatalities at NCB mines, at a rate of 30 per 100,000 men on the books.[37]

Reliable statistics are not available for the period prior to 1851 when the first mines inspectors' reports were compiled, but the inspector for the Midlands, Charles Morton, commented on the considerable technological improvements affecting safety and health which were being increasingly adopted, and suggested a reason:

> Happily, the requirements of humanity and of sound economy are to a certain extent identical as regards improved ventilation and mechanical contrivances in mining, for the latter are conducive, not only to the safety and health of the workpeople, but also to cheapness of production.
> In a well-aired colliery the propwood and timber . . . are preserved longer from decay, and the maintenance and repairs of the underground roads are not so heavy.
> When the supply of fresh air is inadequate, the impurity and heat of the atmosphere incapacitate the colliers and horses from doing a fair day's work, and the operatives are sometimes obliged to leave the pit in the middle of the shift, or to stop away altogether, because the mine . . . is dangerous.

He estimated that labour costs at a badly aired colliery were at least 4*d* a man-day higher than elsewhere; and also stressed the 'great pecuniary loss' to the capitalist of a severe explosion. Elsewhere in the report he says: 'The use of "conductors" clearly tends to economy as well as safety in the drawing of men and material . . .'[38]

A second reason for the improved safety record of the coalmining industry since 1850 is that Acts of Parliament and regulations issued under them have imposed increasingly stringent conditions. The four inspectors appointed in

1850 performed a mainly educational role, but subsequent legislation increased the size of the inspectorate and gave them much greater powers. The comprehensive Acts of 1872, 1887, 1911 and 1954 laid down increasingly detailed rules for the conduct of operations above and below ground, as well as specifying how mines were to be equipped so as to make them safe. Among other things, the Act of 1887 raised the minimum age at which boys could work underground to 12 (raised to 13 under an amending act of 1900); gave the workmen the right to appoint anyone they wished as a checkweighman; and provided for daily supervision of the mine by a person having a statutory certificate. Manager's certificates were introduced in 1872, but under the 1887 Act undermanagers also had to be qualified and a new class of certificate (called the second class certificate of competency) was introduced. Unfortunately, there was nothing to prevent a manager from having charge of a large number of pits and delegating daily supervision to an undermanager, although this was not the intention of the Act. Thus in 1912, 344 men were killed in an explosion at the Hulton No 3 Bank Pit, Lancashire, where the quality of supervision was affected by the fact that the manager had nine separate pits under his control and could only give spasmodic attention to any one of them. The 1911 Act made it illegal for a manager to have charge of more than one mine except with the consent of the Divisional Inspector of Mines, and this was only to be given under certain special circumstances. The 1911 Act also provided for the certification of deputies and surveyors, and similar regulations have since been introduced for engineering staff.

It cannot be denied that some miners' lives have been sacrificed to a short-sighted concentration on production, but the great majority of officials have recognised over the years that bad mining practices which put men's safety at risk also depress production. Under nationalisation, the most productive and most highly mechanised areas have also had consistently the best safety records.[39] The dramatic improvement in safety standards since 1947 has followed partly from the mechanisation of the coal-face operations which has reduced the numbers of men working in the places of greatest risk, but the contribution of safety specialists employed by both the NCB and the NUM, by officials of all grades, by HM Inspectors and by the workmen themselves, should not be underestimated. This great improvement has not created complacency however. The mining community recognises that by comparison with other industries, coalmining is still a hazardous occupation. In a recent report, the Pay Board concluded:

> . . . that the risk of death through industrial accident is higher in coalmining than in any other group of considerable size. Some other groups such as deep sea trawlermen and permanent way staff on the railways show higher overall fatality rates but they are less affected by serious industrial injuries.[40]

5

The Coal Trade

TRANSPORT

In the seventeenth century, most coalmines were very small, worked by one family primarily for their own use, or by a small group of men supplying a local market. There were hundreds of such mines in every coalfield, each adding its mite to the total output.

Where shallow coal seams lay close to a town, or alongside a main highway, somewhat larger units of production existed side by side with the one-man concerns. And in coalfields near the sea, or a navigable waterway, there were even larger units. Much the most important market was London and the South East of England, and the Thames was the natural point of entry for both. Therefore, the collieries of the North East coalfield situated near the sea and the rivers Tyne, Tees and Wear had an overwhelming competitive advantage. Also, they had only one serious competitor, Scotland, in the export trade to Europe. During the course of the seventeenth century, Scottish coal traders benefited by three factors. First, some Scottish coal ('great' coal) was of a quality found especially suitable for domestic use and certain industrial processes; second, taxes on coal exports were easier to evade in Scotland than at Newcastle; and third, English trade was interrupted during the Great Rebellion and continental wars, and Scotland largely made good the deficiency. Scottish exports to the continent were estimated by Nef to have risen from about 1,000 tons a year in the period 1551-60 to 60,000 tons a year (out of a total for England, Scotland and Wales of 150,000) in the period 1681-90.[1] Ayrshire also competed with Lowther's Whitehaven collieries in the Irish market.

Of the three million tons or so of coal produced each year at the close of the seventeenth century, some 1,280,000 was transported by water as Table 5 shows.

John Houghton estimated, in 1693, that the counties which drew their coal through the port of London comprised about 200,000 houses, more than one-sixth of the population of England and Wales, while in the eastern counties supplied with Newcastle coal through other ports (eg Kings Lynn, Boston, Yarmouth), there were about 245,000 houses. Coal was dearer in the South than elsewhere, selling for between four and twenty-five times the pit-head price, and yet consumption was comparatively high. Nef estimates that Londoners burnt coal at the annual rate of 16cwt per head at the end of the seventeenth century, when consumption throughout the country averaged

Table 5 Estimated annual trade in water-borne coal (in tons)

	Destination	1541–50	1681–90
Shipped by sea	The East and South-East coasts of England	22,000	690,000
	Foreign countries and the Colonies	12,000	150,000
	The West and South-West coasts of England (Including Wales)	4,000	80,000
	Ireland		60,000
	Scotland	3,000	50,000
Shipped by river	River Valleys	10,000	250,000
	Total	51,000	1,280,000

Source: Nef, **I,** p 79

only 9cwt a head. The ports of South Wales dominated the Channel Isles, south-west English coast and Southern Ireland trade from the sixteenth century, and by the end of the seventeenth century their exports were running at about 75,000 tons a year.[2]

Towns on the Tyne and Wear had the benefit of cheap coal from the large local sea-sale collieries. Worcester, Gloucester and other towns on the Severn were similarly supplied from Shropshire mines. Coal traffic on the Wye was comparatively light because the small producers of the Forest of Dean were not greatly interested in expanding their market area. Collieries at Wollaton and Strelley in Nottinghamshire had a ready sale for their coal at places like Newark and Gainsborough on the Trent. The Trent was, however, more difficult to navigate than the Severn, and Trentside towns therefore paid higher prices for their coal than towns on the Severn. About 1610, coal valued at 3s 10d the 'rook' at Strelley pit-head was to be sold at 9s 10d, the 'rook', ie almost three times the pit head price, at Newark, some 16 miles down river, while Tewkesbury on the Severn some 50 miles from the pits, was supplied at under 6s a ton in 1678, ie about double the pit head price. Huntingdon Beaumont tried to develop a trade with London in the early seventeenth century but the cost of navigating the Trent made Wollaton prices uncompetitive with coals from Newcastle.[3]

Road transport imposed higher costs than sea or river transport and the only coalfield to have a fairly wide landsale market area was Warwickshire where the Fosse Way and Watling Street cross. The Thick Coal, easily worked, outcropped near the Fosse Way and its low cost of production partly offset its high transport cost to places like Leicester over 15 miles distant. In other coalfields, little coal was carried more than 10 miles by road.[4]

Most landsale coal was carried by packhorses for whom paved tracks were sometimes provided alongside the road in colliery districts. Even main highways were virtually impassable for carts during much of the winter prior to the improvements of the turnpike trusts in the eighteenth century. The bulk of the town merchants' requirements were transported in the autumn, farm carts being used after harvest time for this purpose. At Darley and

Ashover (Derbyshire) tenants had to cart a load of coal for their lord of the manor each autumn in the seventeenth century, and it seems likely that there had been similar obligations elsewhere. The wooden railroad from Strelley to the outskirts of Nottingham, constructed in 1604, was the forerunner of a whole network of rails on Tyneside. Because of the capital costs of constructing them, (estimated at £400 to £600 a mile in the eighteenth century) and need for wayleave, they could only be justified where there was a considerable and reasonably certain demand and a correspondingly large colliery or group of collieries to supply it. The only coalfield which met with these criteria in the seventeenth and early eighteenth centuries was the North East. The only Scottish example known to Duckham earlier than 1750 was a line linking mines at Tranent with the harbour at Port Seton, Cockenzie, opened in 1722. There were, it is true, wooden railways in Shropshire, but these were fundamentally different from the Tyneside ones. They evolved from the narrow gauge lines which brought coal in small wheeled trams from the coal-face to the Severn at drift mines near Coalbrookdale. Even in the Cumberland coalfield, with its similarity to the North East, the first railroad was not constructed until about 1735. It linked the Lowther mines to Whitehaven harbour. The reduction in transport costs was considerable for a large colliery. For example, it was said that waggon transport on the Tanfield line in 1712 was nearly one-and-a-half times the cost of mining the coal, whereas road transport had been three-and-a-quarter times the cost of mining.[5]

These 'Newcastle roads' as they came to be called, supplied coal to the staithes (wharves) on the banks of the Tyne and Wear. Originally, the staithes were built with narrow earth causeways along which the coal was transported by wheelbarrow and tipped into small craft (called 'keels') which carried the coal out to the sea-going ships (the 'colliers'). In the early eighteenth century, piers were built out into the river on piles, and the coal was then loaded into the keels through wooden chutes called 'spouts'. A century later, a device called the 'drop' was introduced to reduce the degradation of the coal caused by shooting it into the keels through spouts. The coal drop had a square frame on pulleys in which the loaded wagon was lowered directly to the keel, with the empty wagon being returned by balance weights.[6]

The keel held eight chaldron wagon loads, ie 21 tons 4cwt.[7] On the Tyne, the round trip from the staithes to the harbour bar at Shields, where the collier brigs were loaded, and back again, took from 12 to 15 hours. The coal had to be shovelled from the keel through the coal ports in the sides of the colliers.[8]

In 1615, there were 400 'sail of ships' employed in the Newcastle coal trade. The average capacity of these vessels shortly after this was 178 tons. By 1710, there were some 600 ships with an average capacity of 206 tons and employing 4,500 seamen, and they carried about 475,000 tons from

Newcastle and 175,000 tons from Sunderland annually.

The Tyne was a fairly shallow river, and the practice of dumping ballast into it made it gradually shallower. However, on the Wear, staithes built out to a point where the coal could be dropped directly into the colliers without the use of keels came into use in the early nineteenth century, the first being opened in 1812. The keelmen, fearing the loss of their livelihoods, rioted in 1815 and wrecked the new staithes.

The largest vessels loaded on the Tyne in the late 1850s had a capacity of only 400 tons, but with the arrival of steam-powered vessels (the first of which carried 650 tons) improvement of the river became imperative, and power to do this was given to a new body of Tyne Commissioners. This body took vigorous and effective steps to deepen the channel, they built piers, river banks, and a new Swing Bridge to replace the old stone bridge; along with these improvements new staithes able to load coal automatically into sea-going vessels came into use. By 1890, the keels had virtually disappeared.[9]

All coalfields benefited from the road improvement of the turnpike trusts which Defoe commented on so enthusiastically during his tour of 1724-6, and from the more thoroughgoing rebuilding which was so marked a feature of the second half of the century. Colliery owners and royalty owners were, indeed, prominent supporters of the turnpike trust movement. For example, the partners in the Butterley Company were among the commissioners for the Derby to Alfreton turnpike, and the company were awarded the contract for making the road. This was completed in 1807 at a cost of £7,286, about £560 a mile, which seems remarkably low by comparison with Wilkes's estimate for an ordinary turnpike road of £1,000 a mile, and for a wide road near a town of from £1,500 to £2,000 a mile. Some turnpike trust toll contractors like Thomas North, on the other hand, invested money in coalmines near the roads which they controlled. A minister in Fife in 1791, while proud of the well-kept turnpike road which served his parish, nevertheless regretted the absence of a canal to reduce still further the cost of coal; while in north Leicestershire the turnpike roads served the smaller producers so well that they had little interest in the ill-starred Charnwood Forest branch of the Leicester Navigation.[10]

The turnpike allowed of an increase in the size of the carts. Up to this time, light carts or wains holding upwards of 12cwt of coal had been common; although according to T. J. Taylor the wains used on Tyneside (of which a large colliery would have as many as 400 or 500) held 17 to 18cwt (which he contrasts with the 42cwt carried in the early railroad wagons which replaced them). A Scottish commentator said that the turnpiking of the Dumfries to Ayr road increased the weight one horse could draw by one-third. Even more important is that turnpike roads facilitated most regular working of the coalmines. By the last third of the eighteenth century, most pits were reasonably well drained, so there was no need to stand idle through winter flooding. Further, there was now greater demand which could only be met by working throughout the year. In the absence of a canal or railway, the alternative to building a road usable in all weathers was to stack coal near the pit-head whenever the roads were impassable, and there was clearly a limit to the capital an owner could afford to have tied up in this way. To take an example, in 1764 Erewash Valley coalowners complained that they had 6,000 tons of coal stacked, waiting for the Belper to Nottingham road to become fit for use.[11]

Canals have attracted the attention of many historians in recent years, and it is unnecessary here to go into detail. The central importance of the canal system to the expansion of the mining industry in the period 1770 to 1830 cannot, however, be over-stressed. Most of the 113 Bills for the construction of canals which were before Parliament between 1770 and 1800 related to mining districts. It is also significant that what is usually regarded as Britain's first true dead water canal, linking the Duke of Bridgewater's Worsley mines with Manchester, was built specifically to carry coal, and had the effect of halving its price in Manchester. This canal, completed in 1761, incorporated some forty-two miles of underground waterway in which coal was conveyed by boat almost from the coal-face. Similar underground canals were constructed in several other mining districts. The reduction in the delivered price of coal following the building of a canal was not always quite so dramatic as in the Bridgewater case. For example, the linking of the Erewash Valley coalfield with Leicester caused a reduction from £1 to 13s 4d a ton. The building of canals sometimes had the effect of increasing the price

Plate 22 Entrance to the underground waterways at Worsley Delph, near Manchester, in 1910. *(NCB)*

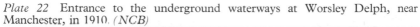

Plate 23 A 'starvationer' canal boat from Worsley being prepared in Brindley's dry dock for its journey to the Lound Hall Mining Museum. *(NCB)*

of coal near the pits because of increased effective demand, while reducing it at a distance.[12] As with the turnpike trusts, colliery and mineral owners were prominent among the canal promoters and investors.

The rapidity with which England became criss-crossed with canals may be contrasted with the Scottish experience. As Duckham says:

> Scotland produced no network of waterways at all comparable with those of the English Midlands or Lancashire and Yorkshire and even the mighty Forth and Clyde carried fairly modest quantities of coal.

The most important coal-carrying canal in Scotland, the Monkland, was projected in 1769 and was supported financially by colliery owners and coal traders as well as prominent members of the Glasgow Corporation, who recognised the advantage it would bring to the city's coal consumers. By 1822, the Monkland was carrying 190,000 tons of coal a year. The Stevenston canal, originating in 1770-2 was only $2\frac{1}{4}$ miles long, but with its five branches, connected a number of pits with Saltcoats harbour.

Although coal was exported from Scottish east-coast ports to the continent and elsewhere, considerable quantities of coal were imported from the Tyne and Wear. In 1815, the imports (approximately 124,000 tons) were not far short of the exports (155,721 tons). This reflects the higher production and transport costs of Scottish coal to which the lack of an efficient system of canals contributed.[13]

With the canals came railway feeders which were usually cheaper to construct than branch 'cuts'. Thus, the Ashby canal (serving the South

Derbyshire coalfield) was originally to have had branch cuts, but the company soon reached the conclusion that tramroads would be much cheaper. The Butterley Company obtained the contract for these lines. Some of the early lines were laid with wooden rails with or without an iron strip to reduce wear. Cast-iron rails, laid on top of the wooden ones, originated at Coalbrookdale in 1767. Farey found two wooden rail roads in Derbyshire in 1808, but by this time, tram plates had become usual in the Midlands. By contrast, many wooden wagonways survived on Tyneside until the 1840s.[14]

One tramway, constructed in 1794-5 principally for the carriage of coal from the Denby district to the Derby canal at Little Eaton and typical of its kind, remained in use until 1908 and a photographic record of it exists. Engineered by Benjamin Outram, it used cast-iron tram plates mounted on massive stone-block sleepers. The wagons had loose superstructures ('boxes') which were deposited into barges by a crane at the canal wharf. A similar crane at Derby lifted the containers out of the barges. With one type of box, commonly called the 'mule', and holding a ton of coal, the crane at Derby deposited it directly on to a two-wheeled dray which then delivered the coal in its container to the domestic or industrial customer concerned. This system had been pioneered by Joseph Butler of Wingerworth (who supplied the original rails for this line) and was adopted by Outram for quite a number of lines besides this one.

The trains (or 'gangs') usually consisted of six to ten wagons drawn by four horses downhill to the canal. On the return journey, four horses could

Plate 24 A tram from the Little Eaton tramroad exhibited at the Lound Hall Mining Museum. *(NCB)*

draw twelve empty or four loaded wagons. There were many colliery railroads where the full wagons ran under gravity, but at Little Eaton the slope was too gentle for this. From the early days, the Little Eaton line faced competition from the Erewash and Cromford canals and the Derby to Alfreton turnpike. After the opening of the Stockton to Darlington railway the Derby Canal Company contemplated the reconstruction of the line or its replacement by a new one so as to improve its competitive position. George Stephenson, who was consulted, estimated that to rebuild the old line, eliminating the worst curves and laying new rails, would cost £13,285, whereas a new line, some eight miles long, would cost £13,250 (exclusive of the cost of land) and give 'a much better communication'. However, neither plan was adopted, and after a temporary resurgence in traffic, the line gradually lost trade to its competitors and was very little used by the time of closure in 1908.[15]

Two even older colliery railroads, leading from the Lane End Pits, Flockton, Yorkshire, to the Calder navigation and dating from the early 1770s can still be traced. On one line are what are reputed to be the oldest railway tunnel and viaduct in Europe, although Dr M. J. T. Lewis considers that they belong to a branch built in the early 1800s.[16]

In 1799, Joseph Wilkes estimated that single railways (or plateways) of this sort cost about £900 to £1,000 a mile, plus expenditure for bridges, culverts, deep cuttings, tunnels and other special features. Fulton, in 1795, estimated the cost at £1,600 a mile while Dr Anderson suggested in 1800 that a double railway in the most favourable situation would cost £1,000, but near London, where wages were high, not less than £3,000 a mile. A subsidiary advantage of a private railroad was that it saved turnpike tolls, but this might be more than offset by wayleave charges. Besides railways built as canal feeders, there were others built to landsale wharves in towns in this period, while Scotland belatedly followed Northumberland and Durham with rails leading to the docks. Dr Anderson estimated the cost of carriage on a railroad eight miles long at 4d per ton against 3s 4d a ton in 'common waggons' on turnpike roads.[17]

It was on colliery lines that steam railway engines, both stationary and locomotive, were first used. Richard Trevithick is usually credited with the design of the first successful locomotive which was tried out on the Penydaren plateway, Merthyr Tydfil in 1804. This, and other early colliery locomotives, had some difficulty from running on cast-iron tram plates. The malleable rolled-iron edge rails first produced by John Birkinshaw of Bedlington Ironworks in 1820 proved much more suitable.[18]

The Stockton to Darlington Railway opened a new era. Its chief promoter was Edward Pease, a prominent local colliery owner (who was a Quaker) and George Stephenson left his enginewright's job at Killingworth Colliery (owned by the 'Grand Allies') to be its engineer. The line was opened in 1825, half of it being laid with cast-iron rails and half with rolled iron. At

LOCOMOTIVE BRADDYLL
This Locomotive Was Built About 1837,
By TIMOTHY HACKWORTH, Who Was
At One Time GEO. STEPHENSON'S Foreman.
It Was One Of Four, Built For The
SOUTH HETTON Colliery. It Finished
It's Life As A Locomotive In 1875,
And Was Used For Many Years
Afterwards As A Snow Plough.

Plate 25 The remains of the colliery locomotive *Braddyll* preserved at the Philadelphia Workshops, Houghton le Spring, Durham. *(NCB)*

first, the motive power was provided by one stationary engine, five locomotives, (four of Stephenson's and one of Robert Wilson's) and horses. In 1827, consideration was given to abandoning the locomotives which had proved far less reliable than horses, but Timothy Hackworth rebuilt and substantially modified the Wilson engine (which he then christened *The Royal George*) and it proved successful. Hackworth engines had a much more powerful blast than earlier designs and they transmitted power 'more directly and efficiently to the driving wheels', by means of cylinders placed one on either side of the boiler.[19]

The Stockton to Darlington Railway, which was the first public railway to use locomotives, enabled the collieries of South Durham to share in the coastwise trade, thus stimulating the expansion of the industry there. Further railways were projected in the district. One of Hair's sketches is of the earliest of these, the Clarence Railway, for which an Act was obtained in 1829. New and improved port facilities (eg Seaham harbour) were provided to handle the coal brought by these railways. The Clarence Railway, which had 47 to 49 miles of line, including branches, is said to have cost double the original estimate of £243,000. The famous competition between rival designs of locomotives, in which Robert Stephenson's *Rocket* performed so well, took place on the Manchester to Liverpool Railway in October 1829, and this new form of propulsion was then firmly established.[20]

The 1820s saw the opening of several new collieries in Leicestershire which relied on turnpike roads to get their coal to market. Because they were in competition with other districts (principally the Erewash Valley and South Derbyshire) served with canals, they were forced to accept narrow profit margins on coal sold more than ten miles or so away. William Stenson, the managing partner of the largest of these collieries, Whitwick, visited Durham in 1828 and was much impressed by the Stockton to Darlington line. He interested John Ellis, a prominent Leicestershire landowner (and, like Pease, a Quaker) in the idea of building a similar railway. Ellis sought the advice of George Stephenson, and Stephenson attended a meeting of interested parties in Leicester in 1829 when he was invited to act as engineer of a line to run from Swannington to Leicester. Of the original capital of £90,000, shares to the value of £58,250 were sold to local people and Stephenson is said to have undertaken to raise the balance in Liverpool. In fact, of the £140,000 which the railway actually cost, Liverpool shareholders subscribed about £40,000.

Following the passing of the Leicester to Swannington Railway Act in 1830, protracted negotiations were held between the Erewash Valley coalowners and the canal companies with a view to reducing the delivered price of Erewash coal at Leicester to meet the competition of the Leicestershire collieries once the railway was completed. An equitable arrangement was, in fact, made (contrary to views expressed by F. S. Williams in 1878 and repeated *ad nauseam* since) and the sales of canal-borne Erewash coal in Leicester increased after the opening of the railway.[21]

A recent study has shown how the colliery owners who were substantial shareholders in the railway were able to negotiate carriage rates with the Leicester and Swannington Company, but this was no longer possible once the Midland Railway Company had taken over the line, as they did for £140,000 in 1845. The Leicestershire owners were not alone in complaining about the monopolistic power of a railway company.

It was as a result of his interest in the Leicester to Swannington Railway that George Stephenson formed a company to sink a large new mine at Snibston in Leicestershire. Snibston, opened in 1832, was the first modern colliery in the district with cast-iron tubbing in the shafts, the cage, tub and guide-rail system of winding, true longwall in place of bank work, efficient furnace ventilation and so on. Its output, which reached about 175,000 tons in the 1860s, was among the largest for a single colliery in the East Midlands. Much of the capital for Snibston was raised among Stephenson's friends in Liverpool, of whom Joseph Sanders and Sir Joshua Walmsley were the most prominent.[22]

In much the same way, George Stephenson's interest in mining coal at Clay Cross, Derbyshire, stemmed from his observations when driving the North Midland Railway tunnel through the district in 1837. The Clay Cross Company which commenced operations shortly afterwards under the superintendence of Charles Binns (George Stephenson's secretary) similarly led the way in technological progress in North Derbyshire and was producing over 300,000 tons a year in the 1860s. Being sunk at the side of the new main line to London, Clay Cross enjoyed a competitive advantage which the company vigorously exploited.

The Swannington to Leicester Railway was typical of many formed in colliery districts to serve local interests and later amalgamated with one of the large networks. In other districts, these larger companies themselves constructed lines through the heart of known coalfields, many of which had been indifferently exploited in the past through insufficiency of transport. Because of the activity of the Midland Railway Company, for example, first the Leen Valley and later the new concealed coalfield of Nottinghamshire were opened up, with the siting of new collieries being largely determined by the lie of the railway line. Throughout the Midlands, the same company was providing landsale collieries with the facility of a widened market; and consequently many were persuaded to expand.[23]

The railway made it possible at last for the inland coalfields to challenge the monopoly of the London market previously enjoyed by the seaborne coal of the North East. At first sight, it seems surprising that no more than a mere trickle of inland coal ever reached London by canal. This is partly due to the fact that some canals were, for a time, prohibited by statute from carrying coal into London. For example, the Grand Junction Canal, opened in 1800, was not permitted to carry coal nearer to London than Grove Park, near Watford, until 1805, when its trade was limited to 50,000 tons on which duty was to be paid.[24]

Tolls of one kind or another were levied on coal entering London from the fourteenth century until 1890. The City Corporation had the right of metage (measuring) the coal until 1831 and a fee was paid for that. The official in charge was called the 'meter', and the men who actually did the work were also called meters. In 1830, there were fifteen principal ship meters and 158 working meters responsible for measuring and certifying the quantities of coal unloaded from ships, and there were also land meters.

Because all coal was metered, it was a convenient medium for raising revenue for both the national and local exchequers; and evasion was difficult, coal being a bulky commodity which could only enter through the port. So from time to time new taxes were levied, some going to the Crown and some to the City Corporation, for the repair of London after the fire, or for discharging the debt on the orphans' fund, or for rebuilding London Bridge, or almost anything else the city fathers could think of. In the reign of Elizabeth I, the tax on all sea-borne coal was 5s per chaldron; and during the Napoleonic War the duty went as high as 9s 4d per London chaldron (ie 7s per ton). In 1830 the duty was 6s per chaldron (4s 6d per ton). Then in 1831, coal duties payable to the national exchequer were abolished and those payable to the City Corporation were reduced to 1s 1d per ton. The same Act abolished the office of meter and the imperial ton replaced the London chaldron as the sales unit. Confusingly the weighmen employed by the trade to weigh coal after 1831 were still referred to as meters. The city dues were not finally abolished until 1890. In 1850, the dues amounted to £176,000. Coal entering London (redefined in 1861 as the City plus the Metropolitan Police District) by canal or rail paid the same dues as sea-borne coal, and markers were erected on canal banks and railway lines where they crossed the boundary.[25]

Early in the century, Farey considered that because of the notably greater cost of canal, as against coast-wise transport, inland coal should be given the benefit of preferential rates of duty, but this was not done. On the contrary, the powerful interests who profited from the coast-wise trade were able to convince the authorities at that time that serious competition from inland coal would affect Britain's national interest by causing a decline in the 'valuable nursery of seamen' which the coastal trade was said to be. [26]

One East Midlands concern which tried to sell canal-borne coal in London was Moira who inserted this notice in *The Times* of 20 September 1815:

> Those who are curious in the truly English blessing of Coals, and give some attention to domestic comfort and economy, would do well at this season to attend to a new quality of Coal found in Leicestershire at Ashby-de-la-Zouch and called Moira Coal, being found on the Estate of the Nobleman of that Title; it is brought by the Canal to Paddington at about 47s per ton or 62s a chaldron, in Boats which carry 44 tons. This appears a little higher than good Newcastle Sea Coals at this cheap time, but on the trial of its economy in use, burning very slow, clear and bright, without the aid of a Poker, without smoke or smell, and having no cinders, it will be found a most agreeable and desirable fuel for the Public Office, the study, bedroom, apartment of the sick, hospital, parlour, and drawing-room, it burns like wood, leaves no more residuum than Charcoal, and requires no attention to keep alight through a long night; the experiment will justify this description. Applications for this Coal by letter only, post paid, to AB at the Gray's Inn Hotel, Holburn, will be attended to if sufficiently numerous and extensive at the price above stated; only the cost at the Pit and Canal freight are included; the carriage of delivery and a tradesman's profit, which on this article would be small, are not considered.[27]

Despite the extraordinary virtues claimed for it, it is doubtful whether Moira coal sold at this price.

There were some who believed that railway transport would be no more successful than canals in enabling inland coal to compete with sea-borne; but George Stephenson expressed a contrary view and he was proved right very quickly. Taking the Butterley Company as an example, they had to wait for a main-line railway until 1847, when the Erewash Valley line was built by the Midland Railway. In 1849, rail sales totalled 62,786 tons; and this rose yearly to 170,588 tons in 1853 of which 45,000 tons was sold to London. By 1860, the company's total rail sale had risen to 491,369 tons compared with canal sales of only 29,531 tons and landsales of 92,116 tons. In 1900, the company disposed of 961,343 tons by rail, 16,733 tons by canal and 52,547 tons by road.[28]

Of the 867,288 tons of coal carried by the Midland Railway from Erewash Valley collieries in 1856, 240,205 tons were received at London stations and a further 105,717 tons at Kew Junction for the South Eastern and South

Western Railway.[29] Rail-borne coal first entered London in 1845, when sales totalled 8,000 tons as against 3,403,000 tons of sea-borne coal. Rail sales rose rapidly, reaching 6,547,000 in 1879 when the sea-borne total was 3,509,000 tons.[30]

But while the coastal coal trade stagnated, the export trade boomed. Export duties, which had been levied at varying rates since the Middle Ages, were reduced in 1834, and abolished in 1850 (although a temporary export tax of 1s a ton was imposed in 1901 and removed in 1906). Britain's coal exports, estimated at 238,000 tons in 1816 and 3,212,000 in 1850, rose to 17,891,000 in 1880, 44,089,000 in 1900 and 73,400,000 tons in 1913. In this field, canals and railways enabled the inland coalfields to participate in the export trade, although South Wales, Scotland and the North East remained dominant. Of the inland coalfields, the one which competed most successfully with the old exporting districts was Yorkshire which had a geographical advantage over the East and West Midlands. Yorkshire is also the only district where a considerable proportion of the vend has continued to be transported by inland waterway. To take a fairly recent year, in 1948, coal for inland consumption, transported by sea direct or by rail and sea, amounted to 1,322,200 tons for Yorkshire, against 16,896,100 for Northumberland, Durham and Cumberland; 3,098,800 for South Wales; 1,996,700 for Scotland and 1,466,000 tons for all other districts combined. Canal and waterway sales totalled 1,615,700 tons for Yorkshire; 931,300 tons for the West Midlands (Staffordshire, Shropshire, and Warwickshire); 812,800 tons for Lancashire; 33,600 tons for the East Midlands (Nottinghamshire, Derbyshire, South Derbyshire and Leicestershire); 9,200 tons for South Wales; and nil for Northumberland, Durham, Cumberland, Kent and Scotland.[31]

For local trade, road transport has remained important throughout, and many small nineteenth century collieries relied entirely on landsales. Clearly, however, collieries served by canals or railways, had a great competitive advantage for sales 10 miles or more from the pits. In Nottinghamshire, one owner, Thomas North, built a private railway system which had, in 1856, 18 miles of standard-gauge rail worked partly by twenty-one fixed and locomotive engines, partly by horses and partly by gravity. This line was used to supply a number of landsale wharves near the main centres of population, some industrial enterprises (eg his own brick works) and the canal system. It was also connected with the Midland Railway. The old established producers in the Erewash Valley, Barber Walker, were worried that North might monopolise the market in Nottingham, and together with Butterley they took steps to prevent this.[32]

Since the building of the public railway network in the middle of the nineteenth century, this has been the dominant mode of transport for coal consumed inland. Thus, in 1950, the railways carried 124·3 million tons of inland coal, against 27 millions by sea, 3·2 millions by canals and waterways and 21·3 millions by road. Since then, larger lorries better able to compete with rail have come into use, but have made only marginal inroads into British Rail's dominance of bulk traffic.[33] Where coal is transported long distances by road this is usually because of physical difficulties on the railways, resulting from shortage of wagons, the closure of branch lines and industrial disputes. The cost of long distance road haulage for bulk traffic is considerably higher than for rail haulage.

MARKETS

The assertion made by Gray in 1649 that:

> Coales in former times was onely used by smiths, and for burning of lime; woods in the south parts of England decaying, and the city of London, and other cities and townes growing populous, made the trade for coale increase yearely . . . .

has been quoted frequently, and expresses much of the truth; although Gray himself mentions elsewhere in his book the importance of coal in the manufacture of salt. The building of chimneys made coal much more acceptable for domestic use in the seventeenth century and this coincided with a disproportionately large increase in the price of wood. Many trades besides lime burning, smithing and salt boiling, adapted their processes so as to be able to burn coal, but nevertheless the greatest expansion in demand was for the domestic market, particularly in London and the southern counties.[34]

The use of the steam engine in the eighteenth century created a new market for small coal, although there was still a considerable unsaleable surplus. The increased demand for coke after Abraham Darby's discovery in 1709 that this could be used successfully in place of charcoal for smelting iron provided another expanding market.[35] In the nineteenth century, the use of steam power in manufacturing industry expanded rapidly and from 1830, railways provided another expanding market for coal both in the production of engines and rail, and in locomotive consumption.

The other great user of coal in the nineteenth century was the gas industry. Before gas was used as an illuminant, the volatile matter given off by coal during carbonisation in open piles or in bee-hive ovens was wasted. The whole purpose of carbonisation had been to produce coke for use in malting, glass making, the smelting or fining of iron or some similar process. Following experimental work done by William Murdoch, Boulton and Watt pioneered gas lighting at their Soho works, Birmingham, at the end of the eighteenth century, and the first public gas light company was authorised by an Act of 1810. This new industry was interested primarily in the gas rather than the coke, and in producing gas they also obtained various by-products. From about 1862, the development of the by-product oven by Pernolet, Coppée, Carvés and Simon gave emphasis to the by-products, on which the

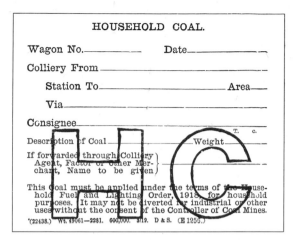

HOUSEHOLD COAL.

Wagon No._____ Date_____

Colliery From_____

    Station To_____ Area___

    Via_____

Consignee_____

Description of Coal_____ Weight___ T. c.

If forwarded through Colliery
Agent, Factor or other Mer-
chant, Name to be given

This coal must be applied under the terms of the House-
hold Fuel and Lighting Order, 1918, for household
purposes. It may not be diverted for industrial or other
uses without the consent of the Controller of Coal Mines.

'(22438.) Wt. 19061—2281. 600,000. 3/19. D & S. (E 1255.)

Plate 26 A wagon label for household coal, 1918. (*NCB*)

modern industrial chemical industry was founded. The by-product works regarded coke and gas as of no more than secondary importance.[36]

The generation of electricity provided another new market for coal from the 1880s, although few coal producers would then have believed that electricity would eventually become a far more important consumer than gas.

It is often convenient to speak of coal as though it were a homogeneous commodity, but this involves a dangerous degree of oversimplification. True, a domestic fire with a good draught will burn any quality of coal (albeit some qualities are far more suitable than others) but most consumers require coal conforming to particular characteristics. Even with modern power stations, which can burn any small coal, it is undesirable that the ash and moisture content should vary outside the limits for which the coal handling and burning equipment were designed, while coal with a high sulphur or chlorine content may damage the boilers or foul the grit arrestors. Some other markets are much more sensitive to quality differences than the domestic or electricity markets. Thus, some coals are totally unsuitable for coking, although the NCB have recently developed a new process in which metallurgical coke of an acceptable quality can be produced from a blend of coking and normally non-coking coals. Three blending plants have been erected at Norwood, Derwenthaugh and Lambton (Durham) at a cost of £1½ million.

The preparation of coal to meet the needs of particular markets has developed greatly in the present century. In the early days of mining, the collier himself prepared the coal by discarding slack, small coal and dirt and sending up only the larger pieces of clean coal. There were many disputes in the nineteenth century over the amounts of slack and dirt sent out with the coal. Deductions were made from wages, tonnage prices were reduced and men were fined or dismissed. Indeed, even in the inter-war period some coalowners discharged men who were caught loading with a shovel instead of

Plate 27 The screening plant built in 1937 by Berrisford Engineering Co Ltd of Stoke on Trent for Ollerton Colliery, Nottinghamshire. *(NCB)*

a screen.[37] Screening plants were introduced at some large collieries in Northumberland and Durham in the late eighteenth century. The earliest screens were arrangements of wooden slats over which the coal was tipped. The small coal passing through the screen was discarded. Later, iron replaced wood, as Ross explained in 1844:

> The corves . . . are drawn to a range of 'screens' consisting of cast iron gratings, about half an inch asunder. Over these the coals are poured, for the purpose of separating the large from the small.

Screening plants became increasingly sophisticated, with picking belts where the dross was discarded and special qualities of house coal were selected, and screen plates with holes of various diameters to sort out the several sizes of coals for their respective markets. From the middle of the nineteenth century, washeries gradually came into use to supplement the screening plants.[38] With machine mining it is impossible to ensure that the coal sent up is clean, and therefore it is necessary to wash a very high proportion of today's product to sort the coal from the dirt.

The changing pattern of inland demand is indicated by Table 6. Since 1938, the locomotive market has completely disappeared and the gas market has dwindled to insignificant proportions. On the other hand, electricity generation now accounts for 53 per cent of total inland consumption.

Coke is another market which has changed considerably. Gas coke, (a by-

Table 6 Coal consumption by uses (expressed as percentages)

	1869	1887	1903	1920	1930	1938
Gasworks	6.7	7.3	9.0	10.0	10.2	10.4
Electricity	—	?	1.8	4.0	5.8	8.5
Railways	3.0	4.8	7.2	7.4	7.7	7.1
Iron and Steel (for Blast Furnaces)	14.8	11.7	10.8	9.6	7.0	6.6
Other Iron and Steel	★	★	★	7.1	4.3	4.0
Coasting Steamers (Bunkers for)	1.3	1.1	1.2	0.9	0.8	0.7
Colliery Consumption	7.1	8.4	9.0	9.3	8.1	6.8
Domestic	22.6	24.7	22.1	22.1	24.0	} 55.9
General Industrial and Miscellaneous	44.5	42.0	29.9	29.9	32.1	

★Included in General Industrial Figures for these years

Source: *Colliery Year Book and Coal Trades Directory* (1951), p579[39]

product of town gas) which can be burnt on open fires, has almost disappeared. On the other hand, the clean air movement has created a growing demand for hard (or metallurgical) coke, which is suitable for closed stoves as well as for smelting iron, and also manufactured smokeless fuels like Coalite, Homefire, and Rexco which are free burning and can be used on open fires.

The export trade has shrunk almost to insignificance since World War II. Exports from the United Kingdom rose from 12 million tons in 1873 to an average of over 65 million tons in the years 1909 and 1913, but thereafter (for reasons we shall consider below) fell continuously to just under 36 million tons in 1938 and little more than one million in 1947. By 1950, they had risen to 13½ million tons but have since fallen to about 2 million tons a year.[40]

MONOPOLY AND COMPETITION

The monopolistic organisation of the hostmen of Tyneside has been previously mentioned. A comparatively small number of substantial producers dominated the market for sea-borne coal, and a cartel was therefore easy to organise. The small independent producers had no way of reaching the London market with their coal and indeed probably had no wish to do so. For them, the local landsale trade was probably sufficient.

In the autumn of 1771, Northumbrian coal producers once more organised a cartel, and this is usually referred to as the 'Vend'. Francis Thompson, a colliery viewer, who claimed credit for initiating the organisation, was appointed secretary. R. W. Brandling explained the operation of the Vend thus:

> ... a Representative is named for each of the Collieries; these Representatives meet together, and from amongst them choose a Committee of nine for the Tyne, and seven for the Wear, which is, I think, the number of Collieries on the Wear: this being done, the Proprietors of the best Coals are called upon to

name the price at which they intend to sell their Coals for the succeeding twelve months; according to this price, the remaining Proprietors fix their prices; this being accomplished, each Colliery is requested to send in a Statement of the different sorts of Coals they raise, and the powers of the Colliery; that is, the quantity that each particular Colliery could raise at full work; and upon these Statements, the Committee assuming an imaginary basis, fix the relative proportions, as to quantity, between all the Collieries, which proportions are observed, whatever quantity the Markets may demand. The Committee then meet once a month, and according to the probable demand of the ensuing month, they issue so much per 1,000 to the different Collieries; that is, if they give me an imaginary basis of 30,000, and my neighbour 20,000, according to the quality of our Coal and our power of raising them in the monthly quantity; if they issue 100 to the 1,000, I raise and sell 3,000 during the month, and my neighbour 2,000; but in fixing the relative quantities, if we take 800,000 chaldrons as the probable demand of the different markets for the year; if the markets should require more, an increased quantity would be given out monthly, so as to raise the annual quantity to meet that demand, were it double the original quantity assumed.[41]

Brandling also believed that the Vend secured for the owners a fair and uniform price for their product and also enabled them

to sell our coals at the port of shipment under our immediate inspection instead of being driven by a fighting trade, to become the carrier of our coals, and to sell them by third persons in the markets to which they are consigned; thereby trusting our interests to those over whom we have no control whatever.

Distrust of the London merchants, who were also monopolists, was to be expected. The London merchants alleged that the limitation of the vend increased prices considerably, and Alderman Waitman 'asserted in 1829, that the tax thus levied by the coal-owners upon the city of London amounted to £800,000'. However, the coalowners maintained that the quota system enabled many 'inferior' collieries to keep open, thus ensuring an adequate supply and helping to hold prices down.

The Vend included owners on the Wear as well as on the Tyne, but the opening of the Stockton to Darlington Railway in 1828 brought competition from the South Durham producers, many of whom did not belong to the Vend. Because of the inroads which non-members made into the market, the Vend had to reduce its prices in 1829. A witness to the Parliamentary committee of 1839, Joseph Pease, MP (a prominent Tyneside colliery owner) put the effect of the limitation of the Vend in perspective when he analysed the final selling price in London in this way: cost of coals free on board at Newcastle, 10s 6d, charges at London coal market, 2s 8d, ship owners' freight charges, 9s 4d and London coal merchants' margin 10s making a total of £1 12s 6d.[42]

The Vend did not control export prices or sales. Therefore, the owner of a heavily capitalised colliery might easily produce well in excess of his quota and export the surplus, possibly at prices considerably lower than those fixed by the Vend for the London market. It was indeed alleged that 'coal was sometimes to be purchased in St Petersburg at half the price of the same coal in the River Thames'.[43]

The Vend was strengthened by an agreement with the coal factors in 1834, but virtually came to an end in 1845 owing to disagreements between small owners and large over quota allocations and to competition from other coalfields. The growth of the railway network broke the North country's domination of the London market, and attempts by the northern colliery owners to bring Midland owners into a price-fixing arrangement were unsuccessful. It may not be without significance that the abandonment of the Vend and the yearly bond occurred together. These were opposite sides of the monopolistic coin, operated by the same association. Similar associations were established in most other coalfields, but not Somerset or the Forest of Dean. There (and also in Shropshire and parts of the West Midlands and Wales) there remained a multiplicity of small mines fairly widely spread, serving local customers mainly by landsale. Even at the close of the nineteenth century, when the railway system was fully developed, some two-thirds of Somerset's output of just under a million tons a year was sold by landsale.[44]

The opening of the canals in the East Midlands created a marketing situation similar to that on Tyneside, and an institution closely modelled on the Vend was fashioned in 1798 in the Erewash Valley to meet the situation. The canal companies, who were concerned to ensure that the full toll was paid on coal shipped, were parties to the arrangement which applied only to canal-borne coal until the railways were constructed, when it was extended to include rail-sales. There were always a few substantial owners who refused to join the association, but small proprietors who tried to remain independent were attacked in various ways. A consultant mining engineer alleged in 1835 that the association was deliberately depressing prices to put the independents out of business, while Barber Walker made it difficult for Thomas North (who was independent for a time) to obtain mineral leases adjacent to his existing workings. Even after joining the combination, North was disliked because his private railway served a number of landsale wharves in Nottingham which enabled him to sell a much higher proportion of landsale coal which was not subject to quota restriction.

With the construction of railways, an association covering a wider area than just the Erewash Valley emerged. A return of rail-and canal-sales of association members for 1860-3 lists eleven in the Erewash Valley (with only three notable non-members: Lord Middleton, Drury-Lowe of Denby and Miller-Mundy of Shipley), seven in Leicestershire and South Derbyshire (including all the substantial firms), but only four in North Derbyshire (Clay Cross, Wingerworth, A. & C. Barnes of Grassmoor and Oakthorpe). Because of competition from the new collieries opening in South Yorkshire and the Leen Valley area of Nottinghamshire, whose owners were anxious to maximise the output of their heavily capitalised collieries,[45] the association's power to influence the level of output and prices was by this time fast slipping away. Towards the end of the century, several attempts were made to revive the price-fixing mechanism of the East Midlands colliery owners, with some of the Yorkshire firms participating, but they met with scant success.[46]

In its hey-day (1802) the Erewash Association was said to have effectively increased the pit-head price of its members' coal from 7s 6d or 8s a ton of 30cwt to 10s 6d a ton of 20cwt. Similarly, an agreement by coalmasters in the Glasgow area in 1776 is said to have forced up the price from 2s 6d to 3s a cart (of 12cwt). A more thorough-going cartel, the Glasgow Coal Company, formed in 1790, operated by allocating an output quota to each of the participating firms and acting as their sole selling agents. New collieries served by the Monkland canal, coming into operation during the war-time boom, broke the monopoly within two or three years although attempts were made to revive it during recessions and in 1813 a new cartel including the Monkland mines was established. As with the Erewash scheme, this body was apparently modelled on the Newcastle Vend. James Dunlop of Govan suggested that a levy should be raised to buy out and close down the least efficient collieries, but this suggestion was not adopted. Unlike the Vend, this Scottish cartel had an export bounty arrangement which helped to boost the trade with Ireland in 1813-14 to 27,820 carts.[47]

This cartel, like its predecessor, soon collapsed. For a cartel to succeed, as in the Erewash Valley over a considerable period, it must be able to control an overwhelming proportion of the output. A non-member offering coal just below the price of the cartel will be able to sell the whole of his product. So long as such independents produce only a small proportion of the total, the cartel's members may be prepared to maintain their organisation's regulations, however irritated they may feel. However, if a substantial proportion of the total output is being sold below the cartel's price, with non-members working full-time to supply the market while the members' pits are standing idle much of the time, then one by one members will be tempted to join the independents.

Duckham notes the informal arrangements between the owners in various parts of Scotland which clearly mirrored similar practices in the English coalfields. Where a colliery owner wished to expand his sale at the expense of others but without precipitating wholesale price competition, several subterfuges were open to him, as with a Staffordshire case of 1797. Lord Dudley's agent, Charles Beaumont, in an attempt to expand canal sales, supplied dealers with 26cwt of coal to the ton instead of the customary 21cwt whilst apparently showing the old weight on the sale documents. A competing owner, F. Dumaresq heard about this and complained that this practice would force other producers to lower their prices and all would suffer, but he was sent an evasive reply.[48] Twentieth-century owners wishing to evade price controls similarly sold best coal surreptitiously at the price of an inferior grade, and there is little doubt that this practice had its nineteenth-century parallels.

The cartels of the twentieth century will be dealt with in a later chapter, but a brief description of an attempted monopoly of the late nineteenth would not be out of place here. This was a scheme proposed by Sir George Elliott, a prominent coalowner, in 1893. He suggested the creation of a trust with which all colliery lessees would be asked to merge. The trust would be 'charged with the working of the coal deposits of the country', and those joining it would exchange their assets for an equivalent share in the trust, one-third as debentures and two-thirds as equity. The initial capital was computed at £110 million. Members of the trust would receive 5 per cent interest on their debentures and at least 10 per cent on their equity. Before a dividend greater than 10 per cent could be paid, Board of Trade approval would be necessary and workmen were to have a share of the excess. A reserve fund was proposed to finance new sinkings and developments, together with an insurance fund for the benefit of workmen and a sinking fund to redeem the capital to render the consolidated property permanent. This scheme did not meet with much support among colliery owners, and nothing came of it. However, the idea was resuscitated in 1919, and more seriously (with the 'Foot Plan') in 1946.[49]

Monopoly practices alleged against powerful colliery owners at various times include drowning neighbouring workings; buying out and closing competing collieries; buying up all the mineral rights in a district so as to prevent interlopers from working them and refusing to grant wayleaves.[50]

Carriers of coal also sometimes acted as monopolists. Thomas North initiated a successful action in Chancery against the Great Northern Railway Company in 1859 because they were acting as dealers in coal though they had no legal powers to do so. North's lawyers pointed out that the railway company's purchases of coal for resale had risen from 73,943 tons in 1850 to 794,332 tons in 1857. They purchased large quantities from particular owners at reduced prices. For example, in 1858 they agreed to purchase all the best house coal produced at Earl Fitzwilliam's Elsecar Colliery at 4s 6d a ton of 21cwt and the contract also gave them the right to buy steam hards for their own consumption at 5s 6d a ton. The average declared value (FOB) of coal exported in 1858 from United Kingdom ports was 9s 2d and one may deduce from this that the average pit-head price was around 5s 6d so the railway company were enjoying a bargain.[51] For Fitzwilliam, the margin of profit must have been small, but he was able to spread his fixed costs, which were considerable, over a large output. North's complaint is summarised in this extract:

> The informant [North] charges that such trading was and is altogether beyond and in excess of the powers of the Company and consequently illegal and that

the effect of it has been and is to enable the Company to control and monopolise the market for sales. The cost of carriage is a very large element in the price charged to the consumer of coals and the power of varying such cost at discretion which the Company in effect have especially when coupled with the facilities in the way of station accommodation, and general service at their command has had the effect of placing freighters at so great a disadvantage as practically to exclude them from the market.[52]

A railway company, having a monopoly of carriage in a particular direction could do a colliery owner a great deal of commercial harm by pursuing a discriminatory policy. For example, the Denaby Main Colliery Company alleged, in 1881, that the Manchester, Sheffield and Lincolnshire Railway Company (MS & L) charged them exactly the same rates for carrying coal to east-coast ports as they did collieries lying 17 miles to the west of Denaby, whereas when Denaby wished to ship coal 1½ miles westwards on the MS & L's line to connect with the Midland Railway, they were charged 1s 4d a ton which virtually priced them out of an extensive inland market.[53]

The railway companies bought most of the canals (eg the MS & L owned the canal which linked Denaby with the east coast) to eliminate competition, and where there were two competing railways, it did not take long for them to agree on a unified rate structure like the Coal Traffic Agreement of 1871 governing rates for carriage to London, made between the Midland and Great Northern companies.[54] Rates fixed as a result of this agreement varied between 4s 8d a ton (0.55d per ton-mile) for the pits nearest London to 8s 7d (0.39d per ton-mile) for the pits furthest from London. Colliery companies co-operated to improve their bargaining position in the field of railway rates, and had much to do with two Acts of Parliament dated 1888 and 1894 which provided machinery for limiting railway carriage charges. However, colliery owners complained that they derived little benefit from this legislation. Indeed, when in 1893 new maximum and minimum rates were specified under the terms of the 1888 Act, the railway companies immediately put up their charges to the maximum. The colliery owners protested vigorously and a Parliamentary committee which was set up as a result recommended changes in the law. Under the terms of the 1894 Act which followed, colliery owners were entitled to apply to the Railway and Canal Commission for a reduction of rates where these were higher than in 1892, but they rarely succeeded. Railway companies resorted to subterfuge to increase their revenue, eg by imposing high terminal and demurrage charges and by withdrawing allowances for degradation.

Colliery owners' associations attempted, by joint negotiations with the railway companies, to secure rate reductions, but again with scant success. The fact is that the railway companies had a much stronger bargaining position than the coalowners. This was generally true also in relation to the supply of steam coal to the railways. One railway could bargain with hundreds of separate colliery owners, some of whom would always be glad of a firm contract for twelve months or more ahead. Usually, the railways carried fairly large stocks which gave them added bargaining strength. Thus, when certain Derbyshire and Yorkshire coalowners combined to try to raise the price of steam hards from 6s 6d to 8s 6d a ton in the summer of 1889, the Great Eastern Railway Company bought very little and, instead of entering into new long-term contracts, ate into their considerable stocks in the anticipation that some owner would give way. This soon happened and the railway company were able to contract for 20,000 tons a year at 7s 3d.[55]

Let us now turn to merchants' monopoly practices. In markets near to a coalfield, the opportunities for middlemen to make excessive profits were limited. Any colliery owner, finding his sale depressed by high prices charged by a merchant, would be able to supply the market direct. Indeed, colliery owners always had a substantial share of the local retail trade. 'Forestalling' of basic commodities was a common enough offence in the Middle Ages and the case of Thomas Marshall of Nottingham who 'forestalled 4 wain-loads of sea coals, not allowing those coals to be led and carried to the King's market in the town aforesaid' may perhaps be taken as typical of many in the fifteenth century.[56] One feature which facilitated the engrossing of the price of coal was the seasonal nature of the trade. Coal was transported mainly in the summer and autumn but the time of heaviest demand was winter. Dealers could therefore buy and stock coal when it was readily available at low prices, then sell in winter when customers were prepared to pay very high prices for it. In 1740, coal was being sold in London at 70s a chaldron against 25s a few months earlier. The difference was a monopoly profit for the merchants who had stocked up in the summer.[57]

The opportunities for forcing up the price of coal by artificially depressing the quantity apparently available were much greater in London than elsewhere. Speaking principally of the London market Nef said:

> It must not be supposed that the nation as a whole reaped the benefit of the low prices at which coal was generally sold at the pits. They worked out chiefly to the advantage of certain middlemen who had mastered the art of buying cheap and selling dear . . . .[58]

In the first half of the seventeenth century, the shippers (owners of the coal ships) bought coal directly from the pits, bargaining for the lowest prices they could get. Arrived in London, the shipper sold his load to a wholesaler called a 'woodmonger'. The Woodmongers' Company had a charter which gave them a monopoly of the right to cart goods through the streets of London. It was because of this monopoly that the wholesale coal merchants joined the Woodmongers' Company. There were a few large consumers of coal who were able to bargain directly with shippers, but even they could not manage without the woodmongers' services. In essence, therefore, the shipper was faced with a monopoly buyer, because the members of the Woodmongers' Company agreed between themselves on the prices they would offer. The woodmongers sold coal to the retailers, and since there were many of these and since they had no alternative source of supply, they were in a comparatively weak bargaining position.

In 1667, the year when the charter of the Woodmongers' Company was withdrawn because of their extortionate dealings, some coal purchased from the pits at 3s a ton or less was sold in London for £4 a ton or more. Every interruption in the coast-wise trade caused by the Dutch wars, or pirates, or storms at sea, forced up the price. In addition, the woodmongers took care to keep much of their stock out of sight so as to create apparent shortages. As Nef reminds us, the materials of the Gunpowder Plot were assembled in a merchant's coal cellar of large proportions. Interestingly enough, the practice of concealing the quantity of coal they had in stock was one of the complaints made against the late nineteenth-century London coal 'ring'.

From about the middle of the seventeenth century, shippers found increasing difficulty in buying coal directly from the pits. Instead, they had to buy from 'fitters', many of whom were formerly hostmen. As we noted in an earlier chapter, hostmen at this time withdrew from the risky business of producing coal to enjoy the certain profits of trading in that commodity. Their monopoly position was based on the fact that they had capital to buy coal for stock and they controlled the keels. Although nominally independent, the keelmen (of whom there were about 1,600 on the Tyne in 1700) were in reality employees of the fitters.[59]

The fitters were reckoned to be bad employers, paying low wages. Equally, they were noted for striking hard bargains with colliery owners. J.C. advises the owner of a new colliery that:

> . . . Fitters have great Interest in loading the Vessels, and by consequence can befriend what Coal Owners they please in the Vend of Coals . . . for the Encouragement of Trade, I am afraid you must abate six Pence per chaldron of the pretended current Price privately . . . .

Only by giving such a rebate, suggests J.C., can a coal owner be sure of finding a fitter to buy it.[60]

Corresponding to the fitter on the Tyne and Wear, another class of middlemen appeared on the Thames. Robbed of their monopoly control of carts after 1667, the coal wholesalers took control instead of the lighters which unloaded the coal ships. Originally, lightermen acted as brokers for the shippers on the coal exchange but from about 1670 they began to trade as principals, buying from the shippers and selling to the wholesalers. Many former members of the Woodmongers' Company adopted the title of lightermen. They were also known as 'crimps' or 'brokers'. In their new role of lightermen, the coal brokers operated a ring, agreeing beforehand on the price they would offer for shiploads of coal. Theoretically, the price of coal was fixed by free bargaining at a coal market (or 'exchange') held each morning at a place called the Room Land at Billingsgate Dock, as Defoe tells us:

> All these coals are bought and sold on this little spot of Room Land, and though sometimes, especially in case of a war, or of contrary winds, a fleet of five hundred to seven hundred sale of ships comes up the river at a time, yet they never want a market.[61]

In practice, many bargains were struck on terms favourable to the lightermen outside the market. Indeed, because they had capital to lend, leading lightermen were able to make contracts with colliery owners and shippers before the coal was loaded or even produced. Such coal as was traded at Billingsgate fetched lower prices than on a free market because of the operation of the 'ring'. The monopoly position of the lightermen was strengthened still further in 1700 when their company was given by charter the sole right to sail barges for carrying goods in the metropolitan area.[62] However, because of the abuses in the coal trade, an Act of Parliament of 1711 made combinations of lightermen illegal. Many of the dealers then adopted different styles ('factors' eventually emerging as the recognised one) although they retained both their control of lighterage and their monopoly practices. Some modern factors are lightermen today. For example, Charrington, Gardner, Lockett & Co Ltd, which has absorbed many firms of eighteenth and nineteenth-century coal factors, have the oldest surviving lightermen's licence (No 8) on the Thames. In 1800 Horne and Davey (since absorbed by Charringtons) owned eighty lighters. Many eighteenth-century lighters were crewed, however, by nominally independent master lightermen and their apprentices who were virtually the employees of the coal factors. A small master lighterman might rise, by a life of industry and frugality, into the ranks of the coal factors; but the very considerable capital needed by someone setting up anew as a factor more often came from some other trade or from land. For example, Benjamin Horne, from whom Charringtons trace their establishment in 1719, was the son of a Quaker glover in Arundel, while John Charrington I was a country gentleman in Surrey before becoming a coal factor in 1790, at about the same time as his second cousin became a brewer. The founder of another firm of coal factors absorbed by Charringtons, Thomas Coote, belonged to a Huntingdon family of corn merchants. He saw coal originally as a profitable return load for his corn carts. Again, Stephenson Clarke & Co trace their origin from Ralph Clarke (1707-85), the son of a Northumberland clergyman who was a coal shipper. His nephew John became a member of the (London) Society of Coal Factors in 1776.[63]

Throughout the eighteenth century, there was an agreement between the fitters on Tyne and Wear and the factors of London whereby colliery owners who wanted to ensure a ready sale for their coals in London paid premiums which the fitters and factors shared.[64]

When Mayhew conducted his monumental survey in the mid-nineteenth century, the factors were still controlling the market; despite the fact that the coal exchange was maintained by the City Corporation as a supposedly free market from 1805: '. . . owing to the combination of the coal factors, no more coals can come into the market than are sufficient to meet the demand without lowering the price.' The number of colliers unloaded in a day was limited, the limit varying with the average price on the previous market day. This arrangement was agreed with the Vend (which now included the owners on the Tees as well as on the Tyne and Wear) in 1834, but was wound up in May 1845.[65]

By this time, however, rail-borne coal had begun to enter the London market and by 1867 rail-and canal-borne coal exceeded sea-borne for the first time. Some inland collieries established their own selling offices in London, almost all appointed travelling salesmen, and most appointed agents. No doubt they were able in this way to sell some coal direct to substantial retail merchants and even to large consumers, but they still needed to sell much of their coal through factors. Factors were often appointed as agents; for example, Mr Hemmings was sole agent for the Duke of Devonshire's Yorkshire collieries in 1838.[66] Factors bought coal for stock, matched customers' needs for particular qualities against coals available, and gave extended credit without which many merchants, and some large industrial consumers, would have found it difficult to manage. Colliery owners rarely had capital available to finance heavy stocking or extended credit. Indeed, a fair amount of capital was provided by factors to their supplying collieries; and in the second half of the nineteenth century, many factors invested in the equity of colliery companies. Factors also controlled much of the transport: railway wagons, barges, ships and road vehicles.

With the tremendous expansion in the output of coal in the second half of the nineteenth century, the number of people employed in selling and distributing it — factors, merchants, dealers, agents, travellers, shippers and the employees of all of them — expanded similarly. Whatever monopolistic practices might remain in London itself, (and attempts were certainly made to maintain the prices of house coal), in the rest of the country the coal trade became reasonably free and open. The expectation of the Shireoaks Company, when they appointed their own traveller on the Great Northern and Great Eastern Railways in 1867, that there would be 'a much larger demand' was no doubt justified. With Denby Drury-Lowe, a much smaller family concern, the Colliery Agent/Manager (Mark Fryar) was his own traveller, visiting customers particularly in the south-western counties. A few domestic customers living in, or near, London, who purchased coal by the wagon-load, were supplied direct from Denby.[67]

The Pinxton Colliery Company in 1894 claimed that they had 'broken through the trammels of the Coal Ring and supply coals direct from their depots without the intervention of middlemen, and their necessarily increased prices'.[68] However, some members of the Sankey Commission in 1919 complained that the London factors' ring was still operating. This allegation was put by R. H. Tawney to Mr W. A. Lee, at that time Secretary of the Coal Mines Department of the Board of Trade (who later became the secretary of the coalowners' organisation, the Mining Association). Lee at first replied, 'I have no reason from my experience to suppose that such combines exist', but later amended this slightly:

> One knows there were working arrangements on a small scale. To speak of any general degree of combination in the coal trade does not in my experience correspond with the facts.

Sidney Webb alleged that one firm (William Cory and Sons), trading under various names, was responsible for a large proportion — possibly as much as three-quarters — of all coal distribution in London. He asked Mr Lee:

> Do you remember the Official Committee that sat to inquire into the coal prices in London . . . you know the evidence given there as to transient combination. They said there was no fixed ring . . . but the price of coal was fixed by half-a-dozen merchants on the Coal Exchange?[69]

A later witness, Mr George Rose who was the chairman of an organisation representing the coal factors, testified that Cory's share of the London wholesale trade was 12 per cent, not 75 per cent as alleged. He also explained that, of the 27,000 to 28,000 coal retailers in the country, only 1,500 to 2,000 bought coal directly from collieries, the rest bought through factors. Although he did not say so, there were three tiers: factors, merchants and dealers. Some dealers bought directly from factors, but many bought through merchants who conducted both wholesale and retail trade. He claimed that the factor's net margin of profit averaged about 7*d* a ton in 1913 and 4½*d* a ton in 1918. The factor's profit on domestic coal was fixed during the War at 1*s* a ton, but the margin on industrial coal was usually lower than this.[70]

According to the chairman of the London Coal Merchants' Society, in August 1914, the average retail price of a ton of Derby best coal in the metropolitan district was 25*s* 6*d*, arrived at as detailed in Table 7. There was an additional, unspecified charge for coal supplied through factors. Presumably this was no more than a shilling.[71]

Various members of the Sankey Commission felt that co-operative societies and municipal enterprises would distribute coal more economically than private traders. As one commissioner, R. H. Tawney, pointed out later, considerable economies were possible in transport which was mainly provided in railway wagons of only 10 or 12 tons capacity, of which about 520,000 belonged to some 10,000 separate owners. He went on:

Table 7 Analysis of retail coal price in London in 1914

	s	d	s	d
Cost at pit	11	6		
Railway rate	6	4		
Wagon hire	1	0	18	10
Loaders' wages		9		
Other wages at coal wharf (eg picking out slate, foremen)		$2\frac{1}{2}$		$11\frac{1}{2}$
Carmen, delivery in big sacks		8		
Driving money and attendance at stables		$2\frac{1}{2}$		$10\frac{1}{2}$
Railway siding rent, demurrage, weighbridge charges, etc		1		1
Sacks		$1\frac{1}{2}$		$1\frac{1}{2}$
Vans, trollies and weigh machines		$1\frac{3}{4}$		
Horse depreciation		1		
Forage and bedding		$5\frac{1}{2}$		
Shoeing, stable expenses, vet, etc		$2\frac{1}{2}$		
Stable rent, rates, heating and lighting		$1\frac{1}{2}$	1	$\frac{1}{4}$
Loss on smalls and weights		4		4
Salaries	1	3		
Establishment charges	1	$1\frac{1}{2}$	2	$4\frac{1}{2}$
Total cost			24	$7\frac{1}{4}$
Profit				$10\frac{3}{4}$
			25	6

> The [Samuel] Commission of 1926 found that the gross profits of the Co-operative Wholesale Society were barely half, its expenses less than two-thirds, and its net profits less than half, those recorded for private traders.[72]

However, accepting that inefficiencies in distribution added something to the selling price of coal, there seems to have been little monopoly profit compared with the pre-railway era, as a comparison of the 1839 and 1914 estimated costs shows.[73]

For the poor customer, buying coal in miniscule quantities, prices were very much higher than the figures normally used indicate. Mayhew in mid-century, described the dealings of small traders called coal-shed men 'who get the coals from the merchant in 7, 14 or 20 tons a time, and retail them from $\frac{1}{4}$cwt upwards'. Still lower down the scale were the street sellers of coals who supplied 'the famishing poor' with coal in quantities as small as 7lb, and who apparently made such substantial profits that many of them rose from the ranks of 'struggling coster-mongers, to be men of substance'.[74] The modern equivalent of the 'coal-shed man' is the small shopkeeper or garage owner selling pre-packed coal in 28lb bags. While this is undoubtedly a convenient way to buy coal, the price is considerably inflated.

Coal factors still operate. They provide long credit for merchants and business consumers, transport facilities, and technical services, and they act as wholesalers, carrying considerable stocks of various grades of coal. Much of the coal they sell is, however, consigned from the collieries direct to their customers. Their profit on such business is very largely a charge for credit. Some factors are also considerable retailers, Charringtons being the largest and best known of these.

The agency agreements which existed in the days of private enterprise were brought to an end when the mines were nationalised. There was no point in paying, say, 6d a ton, to an agent for selling a particular colliery's coal in his territory once all collieries belonged to the same owner.

APPENDIX

Table 8 Consumption of coal (million tons) by main markets 1970-1 to 1972-3

	1970-1	1971-2	1972-3
Power stations	74	68	69
Coke ovens	25	20	23
Industry	18	14	12
Domestic	18	15	15
Other inland markets	13	9	9
Total inland	148	126	128
Exports	3	2	2
Total	151	128	130

Source: *National Coal Board Report and Accounts* (1972-3), p9
1972-3 Reduced to a 52 week year for purposes of comparison

6

Mining Communities

THE MINING VILLAGE

Before the Tudor period, the gradual development of the industry is unlikely to have created many special communities of miners. So far, coal mining was largely a by-occupation of people earning their living from the soil.

From mid-sixteenth century a separate class of pitmen emerged. In a seventeenth-century law case, freehold tenants at Broseley, Shropshire, who worked shallow coal as part of their agricultural activity, complained that James Clifford, Lord of the Manor, had brought full-time miners from a distance to work deeper coal and had built cottages for them on part of the village waste. The newcomers were described by the freeholders as 'the Scums and dreggs of many countries [ie counties] from whence they have been driven'. By the late seventeenth century, there were perhaps 6,000 or more colliers in Northumberland and Durham, and while many still lived cheek-by-jowl with other country dwellers, some undoubtedly lived in separate mining villages. But in most coalfields, colliers were a minority group in the rural community, and remained close to the soil.[1]

The gradual expansion of the industry between, say, 1650 and 1760 was met by the increasing specialisation of mining communities. Sons (and in some areas daughters) inevitably followed fathers into the pits, where once there was a choice of jobs. Mining work became less seasonal, especially after the introduction of the Newcomen engine. Generally the natural increase in the mining population, and its increasing devotion to the needs of the pits, kept pace with the rising demand for labour. Where this increased demand was greatest, (Northumberland, Durham, Cumberland and parts of Scotland) mining communities became increasingly inbred.

J. R. Leifchild in 1856 ascribes the bodily peculiarities of the North country collier to inbreeding. He describes the pitman as diminutive in stature, bow-legged, having long arms, high cheek-bones, and overhanging brows. He considers the effect of their posture at work, but says that this cannot be the determining factor because the collier's shape is already apparent before he starts coal-face work. He then concludes:

> We must look to other causes, in a measure, for an explanation of the bodily defects enumerated above. Pitmen have always lived in communities; they have associated only among themselves; they have thus acquired habits and ideas peculiar to themselves, even their amusements are hereditary and

156

peculiar. They almost invariably intermarry, and it is not uncommon, in their marriages, to comingle the blood of the same family. They have thus transmitted natural and accidental defects through a long series of generations, and may now be regarded in the light of a distinct race of being.[2]

In Scotland, serfdom made inbreeding unavoidable, and so much did colliers become a race apart that they were sometimes refused burial in consecrated ground, according to an old miner giving evidence in 1841.[3]

In the late eighteenth and early nineteenth centuries, new mining communities grew in the Midlands but were generally grafted on to existing villages or small industrial towns. They were nothing like so cut-off from the rest of the working population as the typical North-country mining village so graphically described by two nineteenth-century writers, J. Holland and J. R. Leifchild:

> The pitmen in the north of England reside much less commonly in the towns or villages than in clusters of small houses adjacent to the respective collieries, and forming together little colonies, often more remarkable for the amount of the population, than the cleanness or neatness of their domestic arrangements: the latter circumstance is frequently attributable less to the absence of good housewifery than to other obvious causes. On the other hand it is but justice to remark that many of the houses of the colliers are patterns of cleanliness.

The extract which follows was published in 1856:

> At the old collieries some extremely forbidding dwellings are seen — confined and dismal. In newer collieries they are far better, as at South Hetton . . . . Taken generally, their habitations are mostly in "rows", and these again in pairs; their front doors facing each other, present a space generally clean, unpaved, and without drains or channels. The space between each two rows of back doors, presents along the centre one long ash heap and dung-hill — generally the playground of children in summer, with a coal-heap, and often a pigsty at the side of each door. Each row generally has a large oven, common to all its occupants; there are no conveniences. May not the filthy habits thus engendered, and ingrained as it were, operate in brutalizing the pitmen and their families?[4]

Amazingly, many late eighteenth-and early nineteenth-century mining slums were still being occupied in the 1930s and 40s. Certainly, they had been improved somewhat, but on the other hand, some were more heavily occupied than in an earlier period. Their one advantage was that in Northumberland and Durham they were let by the coalowners virtually rent-free, and in Scotland in 1912, most were let at between £2 10s and £9 a year.[5]

A Liberal Party enquiry of 1925 produced detailed evidence regarding these old houses. The earliest ones in Durham originally had one room and a larder, were built in long rows, had no door or window at the back, and were drained by an open gully running through the yard separating two rows.

Many had been enlarged by building a second storey, connected to the ground floor by a ladder. Some had had yet another small room constructed in the rafters. Generally water was drawn from standpipes in the yard. There were ashpits and earth closets there too.

R. A. Scott James in his appendix to the report, described examples at Framwellgate Moor, Ludworth, Haswell, Consett, and Leadgate, where he:

> . . . saw houses with a single lower room and an upper room approached by a ladder. One that I entered was inhabited by a man, his wife, and six children, of whom the two eldest were girls of 21 and 23 and another was a boy of nearly 18.

Parts of Scotland were even worse as regards overcrowding. Thus:

> In the whole of Lanark county there are 321,471 so-called houses; of these 61,202 consist each of a single room; 155,285 have two rooms. That is to say, two-thirds of the houses are hovells of one or two rooms each . . . .

They were again built in long rows in the stone of the locality, often built back to back, and having floors either of stone or native earth. Some had been built before privies were provided: eighteenth-century colliers used the fields as their privies.[6]

The colliers' rows of the mid-nineteenth century were more solidly built and bigger on the whole than earlier ones, but were similar in design. Most had two rooms downstairs and two up, although some 'double' houses, with a small scullery as well as a kitchen and with a third (attic) bedroom were sometimes provided. Although long rows were usual, occasionally the houses were built in squares. Two particularly good examples dating from the 1840s, Napoleon Square and Holden Square at Cinderhill Colliery, Nottinghamshire, were demolished a few years ago. In this case, and indeed in many others, large gardens with a pigsty were provided. For men who worked much short-time in the summer, these were a boon. Urban miners generally did not have gardens.[7] In South Wales, few houses were built by coalowners. Many were built by private landlords, but there were also building clubs through which many miners bought their own houses in the late nineteenth and early twentieth centuries. While most Welsh houses were better than the Scottish and Durham examples cited above, many were perched precariously on the hill-sides, with the pit head below them and the dirt tip at their back doors.

Many observers commented on the contrast between the inadequate houses of miners and their contents. Leifchild, writing in 1856 about Northumberland and Durham, noted:

> . . . the comparatively showy and costly character of the furniture of the cottages. An eight-day clock, a good chest of drawers, and a fine four-post bedstead — the last two often of mahogany, and sometimes of a very superior kind, were to be commonly observed . . . . The bedding is usually very good even when it is upon the floor of the attic where the younger children of a large family must sleep.

The one real boon of the miner's cottage was the plentiful supply of free or cheap coal allowed in most coalfields. This helped to ensure that there was hot water available for bathing:

> Upon their entrance into their little cottages they proceed to strip and wash themselves, which, from the secluded character of colliery villages, they see no harm in performing somewhat openly; but they haven't much private room. Thus . . . the time of retirement with artizans and mechanics in towns . . . is to colliers the hour of washing. It is as well to know this, when you are passing the pit village at this time, if you have any dislike of soap-suds, which are now repeatedly ejected from the doors.[8]

It is hardly surprising that miners should have chosen to spend so much of their leisure in public houses since they had little room to sit at home. Also, mining is thirsty work; and alcohol provided, as Duckham says, 'relief, escape and relaxation of tensions'.[9]

In the Midlands, beer was taken into the pits and doled out to the men as part payment of wages. Also, where the butty had a financial interest in a tavern, he would expect his men to spend a proportion of their wages there. The annual binding in Northumberland and Durham, colliers' funerals in

Scotland, and the annual fairs and wakes of Lancashire and the Midlands were particular occasions for drunken orgies. Colliers who lived in framework knitting districts were particularly prone to the celebration of 'Saint Monday', and 'feeling Saint Mondayish' is still sometimes used as a synonym for having a hangover. However, by the mid-nineteenth century most colliery managers recognised that drunkenness made poor colliers, and sobriety was thus encouraged. A collier engaged by two Nottinghamshire butties in 1848 was given a sovereign as binding money, but was to forfeit a shilling each time he neglected his work 'through drunkenness or idleness'.[10]

The Methodist revival encouraged temperance (though not total abstinence until the second half of the nineteenth century), and it provided an alternative focus for social life: where once there was only the public house, now there was the chapel too. Where the public house provided robust entertainment: skittles, quoits (in Northumbria), dog-fighting and gambling; the chapel provided penny readings, love feasts and the Pleasant Sunday Afternoon. The public house had an educational and cultural purpose: there, men listened to the paper being read, debated the issues of the day, and conducted trade union business. In the chapel, Sunday schools provided the only education available for most collier children in the first half of the nineteenth century, and for many until 1870. Further, while the culture of the chapel may have been unsophisticated, it was genuine, stimulating a love of music and literature among many.

From about 1840, Methodists became increasingly involved in social work outside the narrow confines of the chapel, taking leading roles in trade unions, friendly (and later co-operative) societies and political movements, particularly the Liberal Party.

Even though chapel-goers were outnumbered by others in most mining villages, their influence was pervasive, leading to better standards of dress, behaviour, and housewifery. But all was not gain. Before the Methodist revival, the North-country miners wore colourful clothes when not at work:

> In their dress the pitmen, singularly enough, often affect to be gaudy, or rather they did so formerly, being fond of clothes of flaring colours. Their holiday waistcoats, called by them 'posey-jackets' were frequently of very curious patterns displaying flowers of various dyes: their stockings mostly of blue, purple, pink, or mixed colours. A great many of them used to have their hair very long . . . when drest in their best attire, it was commonly spread over their shoulders. Some of them wore two or three narrow ribbands round their hats, placed at equal distances, in which it was customary with them to insert . . . primroses or other flowers.

The Staffordshire colliers similarly had worn velveteen holiday clothes decorated with shiny metal buttons, and with very gay worsted garters showing below the knee. Respectable black or navy-blue serge became the standard Sunday and holiday wear of the well-dressed miner after the Methodist revival.[11]

RURAL, URBAN AND NEW VILLAGE ENVIRONMENTS

The miners of the Forest of Dean, Somerset, and Cannock Chase were countrymen. True, some of them lived in colliers' rows near the pits, but even here the clusters of miners' cottages fitted unobtrusively into the rural scene. There were parts of Wales, Leicestershire, Shropshire and Warwickshire of which this was equally true in the nineteenth century. It would be difficult to maintain that these were mining communities at all.

In the case of estate mines, the pit villages were often pockets of paternalism. The Earls Fitzwilliam provided their men at Elsecar with substantial houses with:

> . . . four rooms and a pantry, a small back court, ash-pit, a pig-sty, and garden
> . . . proper conveniences are attached to each six or seven houses . . . gardens
> of 500 yards of ground each, are cultivated with much care. The rent for
> cottage and garden is 2s. a week. Each man can also hire an additional 300 yards
> for potato ground.

In 1845, when this was written, miners' houses elsewhere in Yorkshire were let at 1s 3d to 1s 8d a week, but they were greatly inferior to the Fitzwilliam stone cottages, many of which are still occupied today.[12]

Fitzwilliam also supported schools, a mechanics institute and a library, and he provided pensions for old servants (whether miners or estate labourers) on retirement through ill health or injury and for their widows if they died in service. The usual pension for a widow from 1795 was 2s 6d a week, although this was increased in case of need. The number of pensioners rose from eight (drawing a total of £76 a year) in 1795 to ninety-seven (drawing a total of £673) in 1856. The Fitzwilliams also gave their employees presents at Christmas and St Thomas's day; and financed treats on special occasions. In addition, injured miners in Fitzwilliams' pits were provided with free medical attention, which was by no means universal among colliery owners.

By the standards of the time, the Wentworth schools were efficient teaching units, but they were designed not only to fit the scholars for 'their probable stations in life', but further to inculcate 'habits of subordination, of diligence, and of veneration for the establishments of their country'. The curriculum was therefore restricted to the 'Three Rs' for boys, with the addition of knitting and needlework for the girls. The children of estate workers and miners were taught together. It is significant that the fifth Earl was opposed to any electoral reform which would lead to separate agricultural and manufacturing seats. He believed that there should be a blend of 'agriculturists and manufacturers in the same constituency' just as there was on his estate.[13]

The Earl of Moira's estate was similar in many respects. In 1811, the earl erected two terraces known as Stone Row, Moira, comprising thirty-eight houses with long gardens, each having its own earth closet and ashpit. Each house had a parlour, kitchen, large front room and coal cupboard downstairs and two large bedrooms. The stone outer walls were 18 inches thick and the houses were greatly superior to others in the locality. By contrast, the Whitwick Colliery houses were described in 1842 as 'cramped, dirty, damp and squalid'. Even some of the Moira houses were overcrowded, however. In 1851, the thirty-two houses without lodgers had an average of 5·9 occupants, while the seven with lodgers had an average of 7·6 occupants.[14] A general shortage of houses, and a migratory population made inevitable the sharing of houses in developing mining areas in the second half of the nineteenth century. In hard times some families could not, in any case, afford the rent of a whole house; while a hard-pressed housewife might welcome a lodger to help with the housekeeping expenses, if for no other reason.

Moira was a self-contained community in the early nineteenth century, with the earl paying higher wages than other coalowners, giving pensions to loyal retired employees, paying 'smart money', and helping to finance schools, churches and other institutions. Further, unlike his neighbours, he did not operate the buildas system. In return, he expected and enjoyed personal loyalty from his workmen in contrast to the militant miners of the nearby Swadlincote area who struck in 1842, under the influence of Staffordshire chartists, when their beer allowance was withdrawn.

The Middletons of Wollaton were also paternalistic: they would not employ young boys underground; would not permit any ill-treatment of boys by the butties; and they helped men who were injured at work. And yet, in order to keep down the poor-rate in Wollaton, they would not allow colliers' cottages to be built there. Instead, their growing labour force found accommodation mainly in Radford, a neighbouring 'open' parish which was a centre of the frame-work knitting industry.[15] During the 1844 strike the Chartist stockingers at Radford helped to organise the colliers' strike fund.

The West Yorkshire coal industry had a long history with small 'family' pits whose colliers lived alongside woollen workers in integrated communities. Very different were the mining villages opened up to the East in the second half of the century. Typical of these is Denaby Main which lies between Conisborough and Mexborough.

The Barnsley bed was reached by the sinkers at a depth of 448 yards in September 1867. The Denaby Main Colliery Ltd was formed in 1868 by Pope and Pearson, an established West Yorkshire firm, who brought the nucleus of the original labour force with them. As the enterprise expanded, men were drawn in from many other districts (including Ireland) and were housed in rows typical of the period. The company also provided the schools, church, co-operative store, public house and miners' institute. This was very much a 'frontier' town. Men outnumbered women. For example, in 1871 there were 290 males to 230 females in the eighty-six company houses at New Denaby. Complaints about the boisterous conduct of miners

Plate 29 The first coal raised at Clipstone Colliery, Nottinghamshire, during sinking, 7 April 1922. *(NCB)*

Plate 30 Sinking at Ollerton Colliery in 1924, using the cement injection method. Note the rising main for pumping out the water that is pouring into the shaft bottom as sinking proceeds. *(NCB)*

living in such communities were numerous; gambling, heavy drinking and prostitution being common features. A somewhat later example is Shirebrook in Derbyshire, and even today being 'married according to the Shirebrook rules' is the local euphemism for living 'in sin'.

Denaby has always been particularly militant and recent studies attribute this to its nature as a 'company town' with a population drawn from many districts. This explanation is inadequate, however. While it is true that some other places on the new concealed coalfield have been centres of militancy, it is equally true that most of them have had relatively placid industrial relations.[16] Similar 'frontier town' mining communities with migratory labour, over-crowding and an excess of males, mushroomed in other developing mining areas like South Wales.[17] Two of the most prominent miners' leaders in South Wales, Frank Hodges and A. J. Cook, came originally from Gloucestershire and Somerset respectively.

In Nottinghamshire, the concealed coalfield was developed in two stages. The 'old' concealed coalfield, centred on the Leen Valley, was developed mainly between 1860 and 1880; while the 'new' concealed coalfield further to the east, and centred on Mansfield, was developed mainly between 1890 and 1930. The Leen Valley pits were close to the old mining area of the Erewash Valley, and many men moved only a few miles east to work there. For example, the Booth family moved from Strelley, on the exposed coalfield, to Hucknall, the centre of the Leen Valley, about 80 years ago. One son, Herbert, became a prominent miners' leader and another, Arthur, became a colliery manager. Again, the Ellis family who sank the Hucknall pits came from Leicestershire and many of their men followed them. This was equally true of the Annesley Colliery originally owned by another Leicestershire partnership, the Worswicks. Linby and Newstead also had a large proportion of Leicestershire men, and it was natural that when a colliery at Swannington closed in 1873 and Snibston Colliery similarly closed temporarily in 1882 because the lessors (the Wigston Hospital Charity) refused to reduce the rent when the colliery lessees ran into heavy losses owing to pumping costs and poor quality coal, most of the redundant workmen (about 400 in each case) moved to the Leen Valley.

Additionally, the decline of the frame-work knitting industry coincided with the development of the Leen Valley, so many stockingers were taken into the pits. This was not, therefore, a 'frontier town' situation. It always had about it the feel of a respectable, well-integrated community. And while mining became the dominant industrial activity, there were still many other occupational groups in Hucknall, Bulwell, Arnold and the older villages. There were, it is true, colliery villages built at Bestwood, Annesley and Newstead, but they were close to the older centres of population and were never cut off as the more scattered villages of the new concealed coalfield, with their more cosmopolitan populations, were to be.[18]

Colin Griffin's study of the Leicestershire coalfield has shown that most of the new mines there were manned by men from Leicestershire and adjacent counties. Of 355 heads of mining families shown in the 1861 Census, 53 were born in the same parish as the colliery at which they worked, and 155 more were born either in or just outside Leicestershire and South Derbyshire; 53 others were born in Nottinghamshire, 40 in Derbyshire, 20 in Warwickshire and 12 in Staffordshire. The six who came from Northumberland and Durham were brought to Snibston by the Stephensons. Only eight came from Scotland or Ireland. There was, however, a fairly large Irish population in Whitwick engaged originally in agriculture, and many of the men and boys subsequently became colliers. This was the one 'frontier town' of the coalfield, with much drunkenness and, from 1830 to 1860, rioting between catholics and protestants.

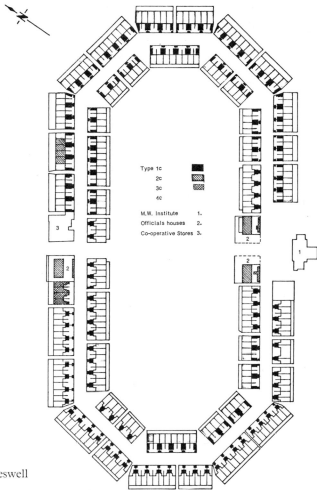

Fig 12   Plan of Creswell
Model Village

In the late nineteenth century, the standards of housing and amenity in new colliery villages greatly improved. Creswell, on the borders of Nottinghamshire and Derbyshire, may be taken as an example. Emerson Bainbridge, MP founded the Bolsover Colliery Company in 1894, taking a ninety-nine year lease of land and minerals at Creswell from the Duke of Portland. Coal turning commenced in 1897, and by 1901 there were 1,400 employees who produced on a day in July of that year a world record output of 3,001 tons. This was a model colliery, well-equipped and well-planned, working virgin top hard coal, and the company built a model village to complement it. There were:

. . . 280 two-storied cottages, built in the form of a double octagon, there being an inner and outer 'circle' . . . . The houses vary in design, and the general effect . . . is very pleasing. The cottages contain five or six rooms and have front and back doors, the latter leading into an enclosed asphalted yard. Within the inner 'circle' is a large green, which is relieved by shrubberies at intervals. A pretty, artistically designed band-stand occupies a position in the centre . . . . There is also a circular playground in which the children can disport themselves.

The company built an institute costing £3,700 with reading, billiards and general recreation rooms and a bar, a library and large lecture hall, and adjoining the institute was a fine cricket ground. Each house had a small flower garden, and tenants could rent 500-square-yard allotments for 4*s* a year. The schools, provided by the Duke of Portland and the company jointly, were managed by a committee representing the owners. The church was erected by the duke; and like the other public buildings, was lit by electricity from the colliery generators. As to sanitary arrangements:

In the yards to the rear of the cottages are enclosed ashpits and outbuildings, and these are cleaned out weekly. The cleansing of the village is done during the night-time, the staff employed for this purpose going on duty at 10.30 and finishing at 5 a.m. The refuse is conveyed away by use of the tram-line . . . an obnoxious smell, however, frequently arises from the field where the sewage is treated.

Fig 13 Type 1c house at Creswell, before and after improvement in 1974

31 The pre-nationalisation pit-head baths changing room at Harworth Colliery, Notts. The clothes were hung on hooks and raised and lowered by ropes and pullies. (NCB)

The primary purpose of the 'tram line', as in many northern mining villages, was to convey the colliers' coal allowance to the houses. Near the model village, some rows of houses were erected by a syndicate; but even so some employees were drawn from the Mansfield district and travelled to work by special 'paddy' trains.[19]

In 1973-4, the model village houses were improved at an average cost per house of £2,977; they now have full central heating, modern kitchens and bathrooms and other facilities.[20] A corresponding environmental improvement is being made by the local authority.

Another East Midlands company, Butterley, provided bathrooms in houses for employees at Kirkby Colliery, Nottinghamshire, as far back as the 1880s, and their New Ollerton village built in the 1920s was also a model of its kind, but there are a number of other good examples of the same period. These model villages, built to serve new mines on the concealed coalfield at which relatively high wages were paid, provided colliery companies with a stable labour force, and despite the odd few contradictory examples, encouraged quiessence among mining communities on the coalfield.[21]

Even for urban mineworkers there is usually a focal point, the miners' welfare institute, around which some community spirit may develop. Most of these have bowls, football, tennis and indoor games teams, and many have brass bands and youth clubs too. People in other occupations who use the welfare develop bonds of sympathy with the miners and to an extent are drawn into the mining community. But in a fairly isolated colliery village, the welfare is central to the life of the community, and is an indispensable institution.

As we have noted, many colliery companies erected institutes for their employees. Nevertheless, the Sankey Commission, recognising that many mining areas were drab, recommended a levy on coal to be devoted to welfare purposes, and this was carried into effect in the Mining Industry Act of 1920. A fund, created by a levy of 1*d* a ton, was to be administered by a body representative of colliery owners and their employees. It was to be spent on pit-head baths, canteens, medical centres, reading rooms, welfare institutes, sports grounds, children's playgrounds, swimming pools, scholarships, and so on. In 1952, the work of the Miners' Welfare Commission was transferred to a new jointly-run body, the Coal Industry Social Welfare Organisation, except that welfare facilities within pit-gates (primarily canteens, pit-head baths and medical centres) became the direct responsibility of the NCB.

The isolation of mining villages is not just a question of distance. Indeed, many of the 'isolated' Northumbrian villages were only a few miles from Durham, Newcastle and Gateshead. Their perceived isolation was due partly to the absence of public transport, and partly to the inward-looking nature of the mining community. The sense of isolation was ameliorated with the growth of public transport systems in the late nineteenth and early twentieth centuries, but has not disappeared.

THE KENT COALFIELD

Kent is a completely concealed coalfield, where coal was discovered in 1890 and production commenced about 1912. When Stanley Jevons wrote, in 1915, he envisaged the development of a major coalfield with ironworks and other new industries, and the growth of 'a hundred villages and many new towns'. In fact, the coalfield has remained small, largely owing to difficult mining conditions.[22] There are only three mines now working.

Since the seams are deep, each colliery sunk in Kent has been fairly large from the outset and has needed to attract a considerable labour force, most of whom have come from other parts of the British Isles. Scottish, Welsh and Yorkshire families predominate in the district. From the outset, comparatively high wages have been paid and superior types of housing provided. Nevertheless, recruitment difficulties remained; and consequently, colliery managements could not afford to be too selective. During the 1920s and 1930s, colliery owners in the older coalfields refused employment to many men they considered to be 'militant'. Some of these, despairing of finding work in areas where they were known, made their way to Kent.

It is perhaps not surprising, therefore, that industrial relations in Kent have been somewhat difficult, nor that the Communist Party should exercise so much influence there. Somewhat paradoxically, there is also a substantial Roman Catholic community in Kent, especially at Snowdown Colliery.

7
The Coal Industry in Decline 1920-70

INTRODUCTION

Between 1850 and 1920, the coal industry expanded rapidly, as the figures in Table 9 indicate.

Table 9 Output, manpower and capital, 1854-1924

	Date	Output (Thousand Tons)	Manpower (Thousands)	No of Mines Producing Coal	Capital Employed
	1854	64,700	214	c3000	c£30m
	1864	92,800	307.5		
	1873-82	138,087	482		
Annual	1883-92	169,922	552		
Averages	1893-1902	203,323	713	3,236	
	1903-12	253,983	936	3,170	
	1913	287,430	1,107	3,024	c£130m
	1920	229,532	1,227	2,571	
	1924				c£200m

Sources: Mineral Statistics 1854 & 1864; Ministry of Fuel and Power, *Statistical Digest;* Samuel Report, QQ5763-4 (Evidence of W.A. Lee)

The fall in output between 1913 and 1920 reflects the one disquieting feature of the period, a massive drop in productivity. Between 1883 and 1913 the output per man per year fell from 333 tons overall (411 tons for underground workers only) to 260 tons overall (322 tons underground). By 1920 there had been a further fall: to 187 and 236 tons respectively. This fall was chiefly due to the operation of diminishing returns to effort which is so much more intense in mining than is often realised, though there were some special factors too, for example, a reduction in working hours and the employment of inexperienced men.

The tendency to diminishing returns can be offset by substantial capital investment coupled with technological innovation. In the life of a particular colliery, productivity will inevitably fall as the faces move away from the pit-bottom and as deeper seams are worked, so that it may appear by looking at the statistics rather than at the reality that increased size is a cause of lower productivity. In fact, increasing the size of the unit is one way of minimising the unavoidable fall in productivity by spreading a larger output over the standard manpower employed on winding, banking, pit-bottom onsetting, etc. In short, when comparing classes of collieries of varying size at a point in time, comparisons of productivity are totally misleading indicators of comparative efficiency, and it is totally untrue to say, as N. K. Buxton does in an article in the *Economic History Review*, that 'there were no *significant* economies to be derived from larger scale mining' between the wars. A small mine worked by three men during the 1926 lockout achieved an output per man-shift (OMS) of 15cwt (equivalent to the national average of 1920) with the most primitive equipment imaginable. But it was only 18 feet deep, and most of the effort was expended directly at the coal-face. Again, in 1945 there were only thirteen collieries in Britain with productivities of 45cwt per man-shift or more, and of these twelve were shallow mines working compact areas of coal with less than twenty men each. They may appear efficient by comparison with large, deep mines, but the comparison is false. A fairer comparison is with the opencast mining of shallow coal in this period, with productivities ranging from 200 tons per man-shift upwards. In the life of a colliery, there is first a period when the coal is near the pit bottom when very high productivities can be achieved. Then, as the faces move away, the labour force and capital invested increase and the overall productivity falls. After a period of years, the owners will probably reconstruct the colliery to work deeper seams, widening and deepening the shafts, installing more powerful winding engines and so on. The productivity will then, all being well, jump to a higher plane, but again, as the faces move away, diminishing returns to effort will be experienced. The most trying time of all is when the only coal left to be worked in a seam is in isolated patches at all points of the compass. Even if the operation is being conducted at the maximum efficiency, productivity inevitably will be low, and costs of production, high.[1] In considering statistics for mines of varying sizes one has to bear in mind, then, that the picture is obscured by the simultaneous operation of increasing returns to scale and diminishing returns to effort.

The fall in productivity between 1883 and 1920 did not spur colliery owners to intensify technological improvement, because an inelastic demand for coal caused its price to rise relative to that of other commodities; and also enabled owners to meet steadily rising wages despite a falling labour performance. (In fairness, coal-face performance improved in some places, but not enough to offset diminishing returns to effort outbye.) This situation came to an end in the winter of 1920-1 when the post-war export boom for British coal collapsed.

The average declared value (FOB) of coal exported from Newcastle, which had been only 12s 4d a ton in 1913 and 26s in 1917, rose to 48s 4d in 1919 and 84s in 1920; it then fell back to 34s 3d in 1921 and 21s in 1922.[2] From then until 1970, the coal industry was declining, although special circumstances in 1923-4 (the French occupation of the Ruhr and an American coal strike) and again in 1936-57 (rearmament, the war-time demand and the post-war boom) masked the process.

Coal was affected, along with industry in general, by the great depression of the inter-war years. Special factors affecting coal were the commissioning of more efficient coal-burning equipment, and increasing competition from lignite, oil and hydro-electric power.

Britain retained her share of the declining world export trade in coal until the mid-'20s. There was then the dislocation of 1925-6 (with Britain's return to the Gold Standard pricing her coal out of some markets, followed by the seven-month-long lockout) and subsequently more effective competition from American and European producers.

Table 10 British coal exports 1913-38

Year	Million Metric Tons	Expressed as a Percentage of World Export Trade in Coal
1913	96	55.2
1924	81	53.3
1929	78	44.1
1934	54	42.5
1938	47	37.6

Source: *Secretary for Mines Report* (1938)

Britain's competitors, faced with difficult trading conditions, invested heavily to improve their performance with the results summarised in Table 11.

Table 11 Productivity trends in European coal industry 1925-36
(Expressed in output per man-shift)

Country	Basic Year	OMS in Basic Year (cwt)	OMS in 1936 (cwt)	Percentage Increase
Poland	1927	23.44	36.20	54
Holland	1925	16.48	25.94	118
The Ruhr	1925	18.62	33.66	81
Britain	1927	20.62	23.54	14

Source: *Reid Report,* p29

Continental bankers were much more willing to invest in heavy industry than their British counterparts, but this was by no means the whole story.

An official committee of mining engineers, the Reid Committee, which reported in 1945, contrasted Britain with the Ruhr. In Britain, there were, in 1924, 2,481 mines producing coal, owned by about 1,400 separate undertakings, whereas in the Ruhr in 1926 fewer than 20 undertakings controlled 90 per cent of the output of 110 million tons:

Concentration of ownership enabled the individual undertakings to close down or merge with others, those mines where conditions were unfavourable and costs high, and to increase the output of the more productive and lower-cost mines.

The mineral rights are owned by the State, and the orderly extraction of the seams is supervised so that regard is paid to the national interest. Each undertaking leases a considerable area of coal and this fact, no doubt, accounts for the large size of the undertakings.

Also, the Ruhr coalowners concentrated their operations on the minimum number of coal-faces; which they reduced from about 13,000 to 4,320 between 1929 and 1933. The overall OMS rose over these years by 23½ per cent without any important provision of coal-face machinery. Instead:

> Much greater integration amongst undertakings in this coalfield has led to the greater exploitation of the most productive seams and important economies have been secured in the effective working time below ground by the . . . provision of mechanical haulage for workmen and shortening of travelling roads.[3]

British coalowners in several coalfields sought to improve their position by forming cartels in 1928. Of these, the Central Collieries Commercial Association (the so-called 'Five Counties Scheme') was the most important, the aggregate output of its members being 100 million tons a year. This cartel had three main features: the available business was shared by a system of output quotas; minimum prices were fixed for the respective classes of coal; and there was an export bounty of 1s 6d to 4s a ton financed by a levy of up to 3d a ton on the total output. The inland coalfields' exports rose from 2·2 million tons in the year ending March 1928 to 5 millions in the following year, although Britain's total exports remained substantially unchanged at around 50 million tons.

Non-members derived most benefit from the cartel, however, because they could work to capacity and sell all they produced by charging prices slightly lower than those fixed by the cartel. Under the Coal Mines Act of 1930, compulsory cartel arrangements, with output quotas and minimum prices, were introduced. Each district had its own cartel, and fixed its own prices. Because of the depressed export demand, the northern coalfields sought to increase their share of the home market by deliberately undercutting the prices of the central districts, with the result that the water-borne share of London's coal trade which had been 42·3 per cent in 1923, rose to 57·7 per cent in 1931. Also, the northern coalfields were able to work nearer to capacity than the central districts. In 1934, for example, coal was wound on an average of 5·5 days in Scotland, 5·27 days in Northumberland, 5·35 in Cumberland and Westmorland, 4·91 in Durham; and, by contrast, 4·12 in South Yorkshire, 3·88 in West Yorkshire, 4·02 in Nottinghamshire, 3·54 in North Derbyshire, 3·28 in South Derbyshire and 3·39 in Leicestershire.

The effect of the quota system was to spread work over efficient and inefficient producers alike, thus stifling major capital investment schemes to increase primary capacity (eg new sinking) which could only pay for themselves if output could by expanded. This was the most important factor in Britain's poor performance relative to continental producers. There was, however, considerable investment in coal-cutting machines, conveyors and gate-end loaders; and on the concealed coalfield of Derbyshire, Nottinghamshire and Yorkshire some large collieries installed man-riding haulages.[4]

Table 12 Output mechanically cut and conveyed

Year	Percentage of Coal cut by machine	Year	Percentage of Coal mechanically conveyed
1898-1902 (Av)	1.6		
1903-12 (Av)	4.9		
1913	8.5		
1920	13.2	1928	11.8
1930	31.1	1930	31.1
1940	63.7	1940	63.7
1950	79.0	1950	84.6

Source: Ministry of Fuel and Power, *Statistical Digest*

Paradoxically, while the quota system helped inefficient producers to keep going, another section of the 1930 Act aimed to stimulate rationalisation schemes. Some voluntary mergers did take place (chiefly large companies buying out competitors so as to be able to close some units, and allocate their quotas to those which remained) but the Act did little to help the process. The Coal Mines Reorganisation Commission, set up under the Act, was replaced in 1938 by the Coal Commission whose chief function was to acquire freehold coal rights on behalf of the State. This process was completed at a cost of £66,450,000 in 1942. The rights attached to former copyhold land remained in private hands however.

During World War II, the Government instituted a system of control which left the organisation of the industry substantially unchanged. The machinery of the cartel was used to control the distribution of coal, and to operate a system of levies and subsidies (the Coal Charges Fund) aimed to keep all pits in production with the efficient subsidising the inefficient. Table 13 summarises the net positions of the respective districts in 1944.

During the war, output fell from 231·3 million tons in 1939 to 203·6 millions in 1942 and 174·7 millions in 1945. This was due partly to the operation of diminishing returns (including, again, a failure to develop new primary capacity) and partly to the loss of 80,000 young mineworkers to the

Table 13 *Coal charges account for 1944*

Net Contributors	AMT per ton		Net Beneficiaries	AMT per ton	
	s	d		s	d
N Derbyshire	1	0.51	Northumberland	0	8.16
Nottinghamshire	1	6.40	Cumberland	10	0.45
S Derbyshire	2	3.82	Durham	3	9.47
Leicestershire	3	7.14	Yorkshire	0	6.29
N Staffs	0	0.03	Lancs & Cheshire	3	4.83
Cannock Chase	0	9.77	N Wales	2	5.10
Staffs & E Worcs	0	10.74	Shropshire	2	5.53
Warwickshire	2	6.59	S Wales & Mon	7	10.16
Scotland	0	6.30	Forest of Dean	3	8.52
			Bristol & Somerset	1	2.82
			Kent	1	11.93

Source: Ministry of Fuel and Power, *Statistical Digest* (1944), pp68-73

forces in 1939-41, but there were other factors: war-weariness, shortages of materials, poor organisation, and indifferent management for example.

The war left the essential organisation of the industry intact, with 746 separate undertakings, 330 of whom produced less than 10,000 tons each in 1943. The Reid Committee concluded that it was:

> not enough simply to recommend technical changes which we believe to be fully practicable, when it is evident to us, as mining engineers, that they cannot be satisfactorily carried through by the Industry organized as it is today.

The colliery owners devised their own scheme of rationalisation (commonly called the 'Foot Plan') but the formation of the National Union of Mineworkers in January 1945 and the election soon after of a Labour Government made nationalisation a certainty.[5]

THE INDUSTRY UNDER NATIONALISATION

The National Coal Board assumed responsibility on 1 January 1947. Its assets included 980 coalmines, in addition to which there were about 400 small mines in private hands producing coal under license from the Board.[6] The Board reported that:

> Taking the coalfields as a whole, many collieries came over to the Board in first-class condition. Many others were in poor shape, and not a few in a pitiable condition.

Compensation for the main assets of collieries was fixed by an independent tribunal at £164,660,000, and this 'global sum' was allocated between colliery companies by expert Valuation Boards appointed by the Minister of Fuel and Power.[7] In addition, the NCB took over freehold coal from the Coal Commission, and the outstanding compensation to former royalty owners, then valued at £78,457,008, became a charge on the Board. Also, additional compensation was paid for subsidiary assets like brickworks; stocks of products and stores; capital expenditure incurred by the former owners between 1 August 1945 and 31 December 1946; and interests under freehold leases.[8]

The NCB inherited a seller's market and this continued until 1956. It is true that many unprofitable mines were kept going in the national interest because the economy needed their coal, and that government control held prices below what the market would bear. It is also true that the losses involved in buying American and Polish coals at world market prices and selling them at our lower controlled prices was a charge against the Board. Nevertheless, annual price increases allowed the Board to balance its books (more or less) despite its failure to improve performance significantly.

This situation came to an end in 1956-7, when the special factors, which had halted the decline in the industry's fortunes during the previous twenty years, no longer applied. Now, to a far greater extent than pre-war, oil made substantial inroads into coal's traditional markets. Crude oil landed in Britain halved in price between 1956 and mid-1970 (at constant money values).[9] To meet this competition, the coal industry had no option but to hold prices down, and this entailed reducing the average cost of production. In 1958, proceeds (average selling price per ton) fell, while from 1961 to 1966 and again from 1967 to 1970, they remained virtually unchanged (see Figure 14).

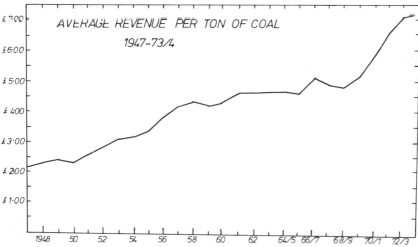

Fig 14   Graph of average revenue per ton of coal, 1947-74

This was achieved by closing grossly uneconomic mines, by concentrating on fewer coal-faces per mine, by reconstructing old collieries, by substantial investment in coal-face and other machinery and by holding wages costs down. The main physical factors are shown in Table 14.

Table 14 Colliery manpower and productivity 1947-73

Year	Average Manpower (Thousands)	No of NCB Mines	Output (Million Tons)	OMS (cwt)	Percentage of Coal Power Loaded
1947	704	958	184	21.5	2.4
1957	704	822	199	24.9	23.0
1968-9	336	317	153	42.5	91.8
1972-3 (53 wks)	264★	281	130	45.8	97.7
1973-4	243★	259	97†	42.3	

Source: National Coal Board, *Annual Reports and Accounts*
★ NB Manpower figures for 1972-3 and 1973-4 are as at the year end
† Affected by a national strike

The capital invested in an average coal-face increased dramatically as power loading machines, armoured flexible conveyers, and increasingly sophisticated hydraulic supports were introduced. By 1972, many face ends had also been mechanised, so that altogether £350,000 or more was invested in a single coal-face. Fewer coal-faces were needed, and fewer men were needed for each face, to produce the required output, as Table 15 indicates.

Table 15 Coal face productivity 1960-73

Year	No of Faces	Av Tons per Face per Day	Face men (Thousands)	Face OMS (cwt)
1960	4.800	158	230	80
1967	1,674	500	140	119
1969	966	590	108	113
1972-3	830	615	★	149

★ NB NCB statistics do not now show face men separately

Output per man per year (all employed) rose from 293 tons in 1950 to 472 tons in 1972-3. However, little new primary capacity was developed between 1965 and 1974, and this inevitably had a depressing effect on overall productivity in the early 1970s.

Throughout the late 1950s and the decade following, wage claims were modest, the union leaders recognising the importance of holding costs down. Consequently, as the Wilberforce Inquiry showed, miners' weekly earnings, which, in October 1960, had ranked third in a table covering twenty-one industries, fell to fifth place in 1968 and twelfth in 1970, rising to ninth after the wage increase of October 1971.[10] Labour costs per ton rose from £1·28 in 1947 to £2·23 in 1957 but then fell, as productivity improved, to £2·05 in 1969. They rose to £5·01 in 1974, but as a proportion of total cost (50 per cent) were lower than in 1947 (62·1 per cent).

The Government, in a White Paper of 1967

> . . . judged it right to base planning on the assumption that regular and competitively priced supplies of oil will continue to be available to us, as they have been in increasing quantities over the past years.

The White Paper also looked for an early increase in electricity generation by nuclear energy, though in a masterpiece of understatement it recognised that 'in a young technology the risk of disappointment must exist'.[11]

In fact, the nuclear power programme is years behind schedule as Table 16, comparing the White Paper estimates with the actual figures, shows.

Table 16 Trends of primary fuel use
(million tons of coal equivalent)

	1957 (Actual)	1966 (Actual)	1970 (Est)	1975 (Est)	1980 (Est)	1970-1 (Actual)	1972-3 (Actual)	1973-4 (Actual)
Coal	212.9	174.7	152	120	80	154	128	121
Oil	36.7	111.7	125	145	160	146	158	155
Nuclear and hydro	1.7	10.2	16	35	90	12	12	12
Natural gas	—	1.1	17	50	70	16	38	38
Total primary fuel use	251.3	297.7	310	350	400	328	336	330

NB Estimates for 1970 and 1975 from 1967 White Paper; estimates for 1980 were prepared, but not officially published, at the same time

By contrast, a report prepared by the Economist Intelligence Unit for the NCB in 1967-8, recommended a national fuel policy which would result in the following 'mix' of primary fuels for 1975:

Table 17

Coal	Oil	Nuclear & Hydro	Natural & Liquified Gas	Total
143.9	145.7	34.5	39.2	363.3

(million tons of coal equivalent)

In the event, the real difficulty for coal in 1974-5 was not on the demand side. With nuclear power so far behind schedule and oil being far dearer than was envisaged in 1967, British coal was once more in great demand but was unable to satisfy the market, so coal for the electricity and coking markets had to be imported. The NCB and the mining unions engaged in a joint production drive to lift deep-mined output for 1974-5 (which, at the June Quarter, was running at an annual rate of 113 million tons) to 120 million tons which was the estimated minimum market requirement. The final result for 1974-5 was 114,652,000 tons from deep mines and 125,120,000 tons in total, including opencast, coal recovered from tips and licensed mines. Since then, the recession in industry has depressed the demand for coal. In the calendar year 1975 total output was 126,650,000 tons and at the year's end there were 30,666,000 tons in distributed and undistributed stocks.

INDUSTRIAL RELATIONS

One of the main differences between owners and the MFGB in the 1920s had been over the level at which negotiations should be conducted. The union favoured national agreements, with an element of national pooling of wages, but the owners successfully insisted on district negotiations.

In 1935, the MFGB persuaded the Mining Association to agree to the establishment of a National Joint Standing Consultative Committee (NJSCC) to discuss matters of common interest, but excluding wages, conditions of employment and things of that kind. On the outbreak of war this body became responsible for stimulating increased production. At the same time, the Mining Association reluctantly accepted the need to have national negotiations on certain wage issues for the duration of the war. Generally, the basic agreements were still made at district and pit levels, but flat-rate additions (to take account mainly of increases in the cost-of-living) were made nationally. These flat rates were paid for by a levy on coal, and, in effect, the poor areas were subsidised by those better off.[12] In 1941, the NJSCC fixed a national minimum wage.

Under nationalisation, the position remained substantially unchanged until 1954 when daywagemen, whose wages had previously been determined in the districts, were assimilated to a national daywage agreement, providing fixed (not minimum) rates. With daywage workers tied to a national agreement, local officials of the union then concentrated on improving the pay of pieceworkers and taskworkers who were still working to local agreements. Consequently, the disparity between piecework and taskwork earnings on the one hand and daywage earnings on the other intensified.

Relationships between NCB and NUM (and other unions) were excellent, and many improvements were made in the miner's lot, one of the most important being the Five Day Week Agreement of 1947. Disputes were usually settled peacefully through the conciliation machinery. The spirit in which this operated is well expressed in an introduction to the booklet on conciliation signed by the leaders of the NCB and NUM in 1966:

> This little book sets out to show that there is no justification whatever for stoppages or strikes, which not only delay the ultimate settlement, but also result in loss of earnings to the workpeople and much harm to the industry. In the strongly competitive field in which coal is engaged losses of production and

productivity are a threat to the livelihood of all men engaged in the industry. The industry cannot afford stoppages or restrictions of work and this booklet has been re-issued at this time to promote the proper use of the conciliation machinery as a contribution to the efficiency of the industry.

During the succeeding five years, there were a number of major changes which produced an entirely different situation.

First, pieceworkers were brought into a national agreement, so bringing about a shift from local to national level in the locus of power. The therapeutic effect of local bargaining was now replaced by frustration especially in high-wage counties like Nottinghamshire which had to accept successive reductions in real wages to allow the low-wage areas to catch up with them.

Second, the era of cheap oil was seen to be coming to an end and this strengthened the bargaining power of the NUM. During the 1960s the leaders of both the NCB (principally Lord Robens) and NUM (Sir Sidney Ford and Will Paynter) had recognised that costs of production had to be held down so as to meet the competition from fuel oil, and this entailed negotiating wage increases which were modest in comparison with many other industries. A great deal of latent resentment was generated among the men, and this showed itself, especially in Yorkshire, in sporadic unofficial strikes in 1969 and 1970.

The tone of the new leadership of the NUM was set by Joe Gormley (elected president in 1971) when he said:

> I can tell you I'm getting off my knees. I have been on them too long. I don't intend to be on them any longer. If acceptance of a high wage policy means pit closures, it means pit closures. So long as we know that and back it up to the hilt, let's go for high wages.

However, this is no indication that Mr Gormley is, in the jargon of the time, a 'militant': he was merely articulating the general frustration of his rank-and-file. Unfortunately, the depth of feeling among the miners and their wives was unrecognised outside the mining community until the strikes of 1972 and 1974 occurred. A detailed discussion of these events would, however, be out of place here.

There seems little doubt that it would be possible for miners to continue to force up wage rates more quickly than industrial workers in general for a short time while the demand for coal is inelastic. But to do so without substantial improvement in productivity would worsen coal's competitive position and intensify the process of substitution. This would lead to a rapid run-down of the coal industry in the later 1970s and 1980s.[13]

Postscript

During the winter of 1970-1, a serious fuel shortage was feared; and coal stocks were being depleted at the rate of a million tons a month, with coking coal in particularly short supply throughout the world. From the middle of 1970, oil prices began to rise substantially as Middle-Eastern oil-producing states realised how little compensation they were receiving for the depletion of their natural resources, so the NCB was able to increase coal prices without fear of losing markets. The Board made an operating profit of £34·1 million and a net surplus after interest of £0·5 million in the financial year 1970-1 compared with an operating profit of £8·8 million and a net deficiency of £26·1 million in the previous year.[1]

Regions making net losses were:

Table 18

Scotland	£9.7 million	Northumberland & Durham	£8.4 million
Yorkshire	£11.4 million	North Western	£3.1 million
South Western	£2.4 million	Kent	£2.1 million

Kent had only three collieries operating, and its operating loss before charging interest was equal to £1·88 per ton.

Regions making net profits were:

East Midlands	£11·4 millions	West Midlands	£3·9 millions

In addition, opencast coal brought in a net profit of £16·7 millions and coal products £5·9 millions, while brickworks made a net loss of £0·3 million.[2]

In the following winter, orders were being lost at the rate of a million tons a month before the seven weeks' strike. One effect of the strike was to weaken the market still further because of a loss of confidence in the NCB's ability to ensure a continuity of supply. Demand, which had been about 150 million tons in 1970-1, fell in 1972-3 to about 120 million tons.

The strike had a disastrous effect on the industry's finances, turning an expected operating profit for 1971-2 of £35 million into an operating loss of £118 million. Adding interest repayments of £39 million and losses from previous years of £34 million, the cumulative deficit at 31 March 1972 was £191 million. To keep the NCB within the limit allowed by law, the government made an immediate grant of £100 million.

This was followed by the Coal Industry Act which sought to put the industry's finances on a sound footing. The various kinds of financial assistance were: a write-off of £475 million of the total accumulated debt to the government of £850 million, thus reducing interest and depreciation charges by £35-£40 million a year; a subsidy to the gas and electricity industries to cover the cost of increased coal burning; and direct support to the NCB of about £120 million a year for three to five years, to cover the costs of coal stocking, higher pensions, a premium on coking coal produced, regional grants and 'social costs' (ie costs arising from redundancies and transfers of employees).[3]

The NCB and the mining unions who had worked together to negotiate assistance on this scale, resolved to ensure the future prosperity of the industry, but this did not prevent another disastrous strike in February 1974, lasting 4 weeks, which necessitated a further programme of assistance. The NCB results for 1973-4, compared with the previous financial year, are summarised in Table 19.

Table 19 NCB results for 1973-4 compared with 1972-3

	1973-4 (52 weeks)	1972-3 (53 weeks)
Saleable output:		
Deep mines	97m tons	130m tons
Opencast	10m tons	11.3m tons
Total	107m tons	141.5m tons
OMS (overall)	42.3 cwt	45.8 cwt
Sales	111m tons	136.5m tons
Operating loss	£112.3m	£39.4m
Loss after interest, etc.	£130.7m	£83.7m
Special Govt Grant	£130.7m	—
Collieries in production at 31 March	259	281

The special government grant to cover the 1973-4 deficiency was additional to grants totalling £128·5 million under the Coal Industry Acts. South Wales made a massive operating loss of £30·4 million (£5·43 per ton) on a turnover of £77·1 million.[4]

Meantime, the coal industry's competitive position had become greatly strengthened by the massive increases in oil prices (which quadrupled during the winter of 1973-4), and interruptions in oil supply. In mid-1974, the 'average price [of British coal] has something like a margin of 30% over oil'. It was anticipated that for the financial year 1974-5 the cost of power-station coal would work out at about $4\frac{1}{2}p$ per therm against 6-7p for oil.[5] This price advantage has been largely eroded by coal price increases in 1975, but will tend to increase once more if producing nations belonging to the

Organisation of Petroleum Exporting Countries (OPEC) extract a larger 'take' from their oil. This rose from $1·75 a barrel in 1972 to $4·30 in October/December 1973. In 1974, the posted price of Saudi Arabian light oil rose to $11·651 a barrel, and its price had risen still further, to $12·376 a barrel by October 1976.[6]

It is probable that oil prices will rise much more slowly during the next few years; indeed, temporary gluts may send prices down for a time as tended to happen on the 'spot' oil market in 1975. But the long-term pressure on resources will inevitably drive up the prices of oil (and other primary fuels) both in absolute terms and relative to the prices of other commodities. World energy consumption trebled between 1900 and 1950, but has almost trebled again to about 8,000 million tons of coal equivalent a year, and could well double again by 1990.[7] The Middle East and Africa, which together consumed only 4 per cent of the world's oil produced in 1971, have 66 per cent of the proved reserves. On the other hand, Japan, to take an extreme case, imports almost 90 per cent of its energy requirements and Western Europe imports about two-thirds. Oil exporting countries are therefore able to exercise monopoly power.[8] Further, many of them are earning depreciating foreign currencies at a rate which they cannot absorb, and it therefore makes sense for them to restrict output so as to conserve a valuable, and appreciating, natural resource.

Britain is fortunate in having natural gas and oil supplies in the North Sea as well as substantial coal reserves. It is estimated that North Sea oil will be producing at between 100 and 150 million tons a year throughout the 1980s, while by 1980 Britain expects to be receiving 6,000 million cubic feet per day of North Sea gas (double the consumption of 1973).[9]

A recent official report suggests that Britain's energy demands will rise by 10 per cent up to 1980 and by a further 10 per cent by 1985. The pattern of supply in 1973 was:

Table 20

Coal	Gas	Nuclear	Hydro	oil	Total
131.3	39.7	9.9	2.0	159.4	342.3

(million tons of coal equivalent)

A tripartite enquiry into the future of the coalmining industry was established by the Government in 1974 and included representatives of Government, NCB and the mining trade unions. It concluded that, taking into account the expected development of oil, natural gas, and nuclear energy, there is likely to be a demand for between 130 and 150 million tons of coal throughout the next decade 'providing the costs of using coal remain competitive with those of other fuels', which is a rather large assumption, unlikely to be realised. However, in June 1974, coal output was running at the rate of only 113 million tons a year from deep mines (ie excluding opencast).

To achieve, and maintain, an annual output of over 130 million tons would require the development of major new projects (like the Selby complex in Yorkshire which is expected to produce 10 million tons a year) and the complete reconstruction of many old mines, which is likely to cost £600 million (at 1974 prices) in addition to the requirement of £70 to £80 million a year for 'ordinary continuing capital expenditure'. Also, the coal industry will need to be able to attract labour; and productivity will need to increase by 4 per cent a year. The closure of permanently and grossly uneconomic mines, most of which are in the peripheral areas, is also unavoidable. In July 1974, discussions began between the NCB and NUM on a new scheme of incentive payments which it was hoped would help to produce 'a new attitude and spirit in the industry',[10] but these failed. Instead a bonus tied to crude national output figures was introduced in March 1975, but proved to be ineffective. There is certainly a long way to go if Britain is to catch up with the Ruhr where many coal-faces produce well over 2,000 tons a day. The average daily output per face (DOF) in West Germany rose from 834 tonnes (metric tons) in 1969 to 1,197 tonnes in 1973, while the United Kingdom's average rose only from 575 tonnes to 577 tonnes. In 1964, face productivity in Britain (181 man-shifts per thousand tonnes) was marginally better than in West Germany (189 man-shifts per thousand tonnes) but by 1973 the West Germans had shown a much greater improvement (to 99 man-shifts) than the United Kingdom (125 man-shifts). It is true that the seams worked in West Germany are flatter, thicker (average section 68 inches compared with 57 inches in Britain) and the coal is not so hard; but this would not account for the better productivity elsewhere underground and on the surface. In 1973, to handle a thousand tonnes outbye the coal-face took 132 man-shifts in West Germany and 170 in Britain, and there was a similar advantage on the surface: 50 man-shifts in West Germany and 92 in Britain. The difference is accounted for largely by the considerably larger size of the average colliery in West Germany, giving economies of scale.[11]

Few countries are in Britain's fortunate position as to indigenous sources of energy. But for the world as a whole, rising oil prices inevitably stimulate the process of substitution. For example, one authority estimated that when the price of oil (FOB Persian Gulf) reached $4·00 to $6·50 a barrel (in 1973 prices) Canadian tar sands and US oil shares would become commercial propositions. In fact, these prices were reached in late 1973; and during the first half of 1974 the average price was about $9 to $9·50 a barrel. It seems likely that there will be substantial production of oil from these environmentally damaging sources by about 1980.[12]

Substitution also takes the form of improved insulation and improvements in the efficiency of energy-consuming equipment. There is tremendous scope for saving here. Another substitute, nuclear energy, unfortunately poses environmental hazards of such magnitude that, by comparison, the environmental damage caused by fossil fuels is insignificant. In 1974, the Secretary of State for Energy ruled against the potentially dangerous American Light Water Reactors which the CEGB wished to build, and instead proposed a relatively cautious development programme based on the Steam Generating Heavy Water Reactor (SGHWR) of which a 100 megawatt (MW) prototype has been operating at Winfrith since 1968. Additionally, the 250MW experimental Fast Breeder at Dounreay was to be 'brought to full power'. Further consideration is now (October 1976) to be given to this matter.

Looking to the longer term, there are only finite quantities of fossil fuels (and, for that matter, Uranium) in the Earth's crust. New sources of energy will therefore be needed. The Energy Technology Support Unit established at Harwell by the Secretary of State for Energy 'is assessing the feasibility of producing energy on a significant scale from solar, wind and geothermal sources'and is conducting a £1 million 'wave power feasibility study'.[13] Similar experimental work, but on a larger scale, is being conducted in America and Russia. Tidal energy may also be more fully exploited in the future.

One source of energy which remains remarkably under-exploited is methane. This is produced in sewage works, some of which use it to generate electricity, but much goes to waste. Most colliery methane is also wasted. It is perfectly feasible for methane (and methanol) to be deliberately produced from vegetable matter as a substitute for petroleum products. Many farms could manufacture their own fuel at minimal running cost; and there is no intrinsic reason why large-scale commercial ventures based, for example, on the harvesting of tropical rain forests or weeds grown on lakes, should not be developed. To discuss this in depth would, however, be beyond the scope of the present volume.

The financial position of the British coalmining industry in November 1976 is worrying. The expected upturn in the economy, which it was hoped would reduce coal stocks in the Autumn, has so far hardly made itself felt. Further, as an NCB publication issued to colliery officials has noted:

> Output and productivity are running consistently below last year's level. The half year's deep mined results were: cumulative productivity 42·4 cwt (1·2 cwt below a year ago); face productivity 149·6 cwt (4·3 cwt down); output 49·3 million tons (3·4 million tons down).

The Coal Industry National Consultative Council has set up a Performance Improvement Team, under the chairmanship of Mr John Mills, Board member for mining, with members drawn from management and the three mining unions 'to see how studies in the coalfield and at the collieries can help reverse the downward trend'.

Unless a significant improvement in productivity can be achieved fairly soon, the *Plan for Coal*, providing for the expansion of the industry in the 1980s, will inevitably be endangered. Accordingly, the tripartite committee, representative of the Government, the NCB and the mining unions, which endorsed the *Plan for Coal* has been reconvened to study the situation and suggest remedies.[14]

Looking to the longer term, the NCB and the mining unions have recently set their sights on an annual output target of 150 million tons from deep mines plus up to 20 million tons from opencast by the year 2000. On present trends, however, our deep-mined output in 1977-8 will fall below 100 million tons for the first time (disputes apart) since the 1860s. Further, the stagnation in productivity, coupled with the increased costs entailed by a reduction in the retirement age of miners currently being negotiated, will seriously worsen coal's competitive position; and this poses a final question. Is the resurgence in the coalmining industry's fortunes since 1970 permanent, or will it be seen in the future to have been no more than a temporary amelioration of the long-term tendency to decline?[15]

STATISTICAL APPENDIX

Table 21 Analysis of average costs for all NCB mines, 1973

	Deep Mines £ per ton	Opencast Mines £ per ton
Wages (including allowances in kind)	3.02	0.01
Wages charges	0.75	0.00
Materials and repairs	1.64	0.01
Power, heat and light	0.31	0.00
Salaries and related expenses	0.36	0.02
Other operating expenses	0.54	0.32
Overheads and services	0.57	0.36
Depreciation	0.47	0.06
Paid to opencast contractors	NA	4.63
Total Average cost per ton	7.66	5.41
Operating profit/loss	0.68 (loss)	1.92 (profit)

Source: National Coal Board, *Annual Report*

Table 22 Inland energy consumption on a primary fuel input basis
(Million tons of coal or coal equivalent)

Year	Total	Coal	Petroleum	Natural Gas	Nuclear electricity	Hydro-electricity
1969	318.5	161.1	135.7	9.2	10.5	2.0
1970	329.6	154.4	145.6	17.6	9.4	2.6
1971	325.9	138.7	147.3	28.4	9.7	1.8
1972	331.3	120.9	157.6	40.3	10.5	2.0
1973	346.1	131.3	159.4	43.5	9.9	2.0
1974	331.1	116.0	149.0	52.1	11.9	2.1
1975*	320.5	120.3	132.9	54.5	10.8	2.0

* Figures for 1975 are provisional

Table 23 Coal supply (Thousand Tons)

Year	Net inland supply	Total[1]	Deep mined	Opencast	Imports[2]	Exports[2]	Net Imports
1970	141,560	144,791	134,526	7,760	78	3,309	−3,231
1971	148,592	147,081	134,322	10,498	4,164	2,653	+1,511
1972	123,124	119,927	107,836	9,815	4,918	1,721	+3,197
1973	128,897	129,906	118,140	9,964	1,642	2,651	−1,009
1974	110,362	108,707	98,413	9,085	3,491	1,836	+1,655
1975	129,505	126,650	115,557	10,249	5,003	2,148	+2,855

[1] Includes an estimate for slurry, etc, recovered and disposed of otherwise than by the NCB
[2] As recorded in the Overseas Trade Statistics of the United Kingdom

Table 24 Colliery manpower and productivity at NCB mines

Year	Number of wage-earners on colliery books at end of period (Thousands)		Absence (Percentage)	Average output per manshift (cwt)
	Total	Underground	Total	Overall
1969	300	235	18.3	43.45
1970	283	221	19.8	44.10
1971	279	218	18.1	43.94
1972	266	210	16.6	43.78
1973	245	193	18.0	45.19
1974	246	194	16.3	42.83
1975	245	194	16.2	44.92

Table 25 Inland energy consumption on a heat supplied basis (Million therms)

Year	Total Primary Energy	Used by fuel industries and losses in distribution	Final consumption						
			Total	Coal	Other Solid Fuels	Other Coal-derived Fuels*	Petroleum	Gas†	Electricity
1969	81,363	24,595	56,768	12,883	6,409	641	25,628	4,962	6,245
1970	83,481	25,528	57,953	11,839	6,008	621	27,198	5,720	6,567
1971	82,521	25,519	57,002	9,867	4,949	542	27,617	7,270	6,757
1972	83,847	25,807	58,040	8,085	4,612	513	28,634	9,193	7,003
1973	87,850	26,847	61,003	8,064	4,743	586	29,623	10,495	7,492
1974	83,494	25,006	58,488	7,519	4,321	437	27,193	11,736	7,282
1975	80,584	24,747	55,837	6,440	3,611	460	25,744	12,309	7,273

* Coke oven gas, creosote/pitch mixtures and other liquid fuels derived from coal
† Town gas and natural gas supplied direct   NB The heavy usage by the fuel industries themselves

Table 26 Inland consumption of coal (thousand tons)

Year	Total Inland Consumption	Collieries	Publicly Owned Power Stations	Gas Works	Coke Ovens	Other Conversion Industries	Industry	House Coal	Colliery disposals	
									Anthracite and Dry Steam Coal	Miscellaneous
1969	161,159	2,032	75,883	6,867	25,373	3,846	21,367	19,745	1,883	4,163
1970	154,407	1,886	76,017	4,212	24,940	4,084	19,304	17,882	1,990	4,092
1971	138,705	1,556	71,696	1,826	23,182	4,406	15,581	15,366	1,621	3,471
1972	120,945	1,383	65,611	566	20,152	4,475	11,478	12,569	1,756	2,952
1973	131,288	1,359	75,628	503	21,543	3,550	11,890	12,519	1,755	2,540
1974	116,043	1,236	65,967	105	18,169	3,728	10,902	11,848	1,623	2,465
1975	120,304	1,218	73,391	10	18,783	3,999	9,532	9,844	1,609	1,918

Source of Tables 22-6: Department of Energy, *Energy Trends, A Statistical Bulletin*, March 1975 and April 1976

Glossary

Adit A tunnel driven into a hillside in connection with mineral working, for transport, ventilation or drainage (or all three).

After-damp A mixture of carbon monoxide and other gases resulting from an explosion or fire.

Agent An official senior to a colliery manager, usually responsible directly to the owner.

Air crossing Where intake and return airways cross, they have to be kept separate by taking one, usually the return, over (like a bridge) or under (like a tunnel) the other.

Backbye See inbye.

Back-shift The afternoon-shift; sometimes used to mean either afternoon-or night-shift.

Bait See snap.

Bank The colliery surface; coal-faces are also sometimes called banks (or benks).

Bank work A primitive system of long-way working practised in Yorkshire, Derbyshire, Nottinghamshire and Leicestershire mainly between mid-eighteenth and mid-nineteenth centuries.

Banksman The man who operates shaft signals at the pit-top.

Bantle Formerly, the men riding the rope on one draw. Now applied also to a cage-load of materials.

Barrier A solid pillar of coal left usually to prevent the migration of water or gas from old workings.

Barrowman A synonym for putter (qv) or haulage hand used mainly in the North East.

Basset The outcrop of a seam at the surface.

Beat knee Bursitis (swelling of the bursa) caused by kneeling. Sometimes called 'housemaid's knee'.

Bearmouth In Cumberland, a drift mine.

Bell pit A shallow shaft, where coal is worked around the pit-bottom until the sides are in danger of collapsing when it is abandoned. Seen in section, it has the general shape of a bell.

Big butty system Where a pit was let to one (or several) superior workmen who provided the circulating capital, paid the men and effectively managed the works, being paid so much per ton.

Blackdamp A mixture of nitrogen and carbon dioxide and found generally in coalmines, resulting from the oxidation of coal and timber. Its production may leave insufficient oxygen to support life.

Blower A violent discharge of firedamp.

Bond The document by which miners were bound to their employers for a year at a time in Northumberland and Durham (and Scotland too after the abolition of serfdom).

Bord-and-pillar work North Eastern variant of stall-and-pillar. The bords were rectangular excavations separated by solid pillars of coal left to support the roof. In the nineteenth century, the pillars were also largely removed in a second working.

Bord gate Term used after 1840 in the East Midlands for a main roadway, not to be confused with bord-and-pillar.

Boring Drilling holes through the strata from the surface by bore rods to establish whether coal is present was pioneered by Huntingdon Beaumont about 1600. This term is also used in modern mining practice for the drilling of shot-holes.

Brattice Where a colliery had only one shaft, this was divided from top to bottom by a vertical wooden partition called a 'brattice'. Air flowed down one segment of the shaft and up the other. Brattices, nowadays usually of cloth, are similarly used in headings to facilitate a flow of air.

Broken working In stall-and-pillar, the working of the pillars.

Buildas system Payment of wages partly in beer, common in parts of the Midlands prior to mid-nineteenth century.

Bull week The week before a holiday when men maximised their earnings by making full time and (in piecework) boosting their output.

Butty In the Midlands and North Wales, a sub-contractor.

Cage A steel platform with sides and a roof in which men, materials and mineral are drawn up and down the shaft.

Cannel coal A shiny coal which gives a good light when burning (hence 'cannel' – thought to be a corruption of candle).

Cavilling system Allotting working places periodically by the drawing of lots practised in Northumberland and Durham.

Chaldron The principal capacity measure used in the North Eastern coal trade. The Newcastle chaldron was equal to 53cwt and the London chaldron about 27cwt.

Charter master Staffordshire name for a big butty.

Checkweighman A man who was employed by the workmen paid by piece work to check the weights recorded at the pit-top weighbridge.

Chokedamp A synonym for blackdamp. Many nineteenth century writers applied this term also to afterdamp, and this has confused some present day historians.

Cleat The lines of cleavage of coal, analogous to the grain of timber.

Coal preparation plant The screens (where coal is separated into different sizes mechanically) and washery (where coal is separated from dirt).

Conductors See guides.

Consideration Extra payments made to pieceworkers supposedly to compensate for abnormal working conditions.

Corf A hazel basket in which coal was conveyed from the coal face to the surface before the introduction of wheeled tubs and cages.

Corporal Man in charge of the underground haulage hands (called a 'doggy' in Staffordshire.)

Crab Capstan used for lowering heavy objects (eg pump rods) in shafts. They can be either hand, horse or steam operated.

Cracket Stool, usually three-legged, used in some districts by hewers either to sit on or to support their shoulders.

Creep The heaving up of the floor of the roadways underground, otherwise called 'floor lift'.

Crib or curb Segmental wooden rings in shafts on which tubbing or brickwork rests.

Crook An iron hook on a chain used to fasten the 'guss' (hempen harness worn by boys) to the 'put' (box of coal) in Somerset.

Crush Convergence of the strata.

Cupola Ventilation furnace chimney, sometimes called a 'cube' or 'tube'.

Darg In Scotland, a day's hewing task, elsewhere called a 'stint'.

Datallers See shifters.

Davy Safety lamp invented by Sir Humphry Davy in 1815.

Day-hole A drift mine.

Deck The floor (or floors) of a cage. Originally cages were single deck, but double- and treble-deck cages became common.

Deputy Colliery official inferior to the overman, responsible for the safety of the men under his charge.

Detaching hook A device by which an overwound cage is detached from the rope and is held firmly in the headgear.

Dip Coal seams are almost always inclined. Working down-hill is said to be to the 'dip', and working up hill to be to the 'rise'. The 'strike' of the seam is its level course (at right angles to its inclination or slope).

Doggy See corporal.

Doors Used to control the flow of air underground.

Double working In bord-and-pillar, where two men worked in a bord. An extra payment was made for the inconvenience.

Downcast The shaft down which the air flows.

Draw Each passage of the cage through the shaft is called a draw.

Drift A heading. A drift mine is one driven from the surface (as opposed to a shaft mine).

Dumb drift In furnace ventilation, a tunnel isolated from the furnace carrying the return air into the upcast shaft over the top of the furnace, so as to reduce the risk of explosion.

Engine In old mining terminology, 'engine' usually meant pumping engine, whether powered by horse, water or steam, so a pit from which water was pumped was an engine pit.

Face The working face of coal.

Fall A fall of roof at the coal-face or in a roadway.

Fault A fracture of a coal seam caused by earth movement; is called an up-throw (where the seam continues at a higher level) or a down-throw (where it continues at a lower level).

Filler A man employed on filling coal at the face.

Fire bucket or lamp Bucket like a night-watchman's brazier placed or suspended in an upcast shaft to aid ventilation.

Firedamp Inflammable gas whose chief constituent is methane. It is emitted from the seam (and also the floor and roof) during working.

Flat In bord-and-pillar, the place where the corves were lifted by crane on to the rolley.

Footrail, footrill or footridge A drift mine.

Fore-shift In Northumberland and Durham, the early hewing shift.

Gate (gateway or gateroad) Underground roadways, a term used mainly in the Midlands.

Getter A man producing coal at the face. In many cases he was also the filler (qv).

Geordy The safety lamp invented by George Stephenson in 1815.

Gin See whim-gin.

Glenny A safety lamp; probably a corruption of Clanny, after Dr W. R. Clanny who invented several lamps, the first in 1813.

Goaf (or gob) The waste area from which coal has been removed; is partly filled with small coal and debris.

Gob fire Spontaneous combustion of small coal in the gob.

Gob road Gate supported by pack walls for bringing coal from the face in hand-got longwall working.

Grove A drift mine.

Guides Rails or ropes running vertically down the shaft to hold the cage steady. Also called conductors. The slides fitted to the sides of the cage which hold it firm to the guides are called 'shoes'.

Hanger-on (or hooker-on) The man who used to hang corves on the rope in the pit-bottom. Also called an 'onsetter' (a term still used).

Headstocks (or headgear) Frame with pulley wheels over a shaft.

Heapstead The building over the pit-mouth.

Hewer A coal-getter.

Hoggers Stockings without feet formerly worn by miners in the North East.

Holing Undercutting the face of coal, (hence 'holer', the man carrying out this operation).

Inbye Towards the coal-face (as opposed to outbye or backbye - away from the face).

Intake Airway along which fresh air is taken into the workings, as opposed to the 'return' airway carrying foul air away from the workings to the upcast shaft.

Jack-roll A hand winch.

Jig A self-acting inclined plane.

Jumper A tapered round iron bar with a sharp point formerly used for drilling shot holes manually, especially in the Midlands.

Keeker In Northumberland and Durham, a surface foreman.

Keel Boat holding about 21 tons of coal formerly used on the Tyne and Wear for carrying coal from the staithes to the collier brigs.

Kenner Shouted at the end of the shift meaning 'stop work'. A Midland synonym is 'loose-all'.

Keps Catches on which the cage rests at the pit top. When the cage is to descend, the banksman releases the keps.

Kerf (or kirf) The cut taken along the bottom of the coal-face originally by a man with a pick (the 'holer') and later by machine. Hand holing produces far more slack than machine holing, so cutting machines were more readily adopted where seams were thin.

Kist In Northumberland and Durham a deputy's tool chest.

Laid out A corf or tub of coals containing an excessive quantity of dirt, for which the hewer was fined.

Landsale Coal sold for transport by road (as distinct from sea-sale, canal-sale or rail-sale.)

Layering Firedamp, being lighter than air, sometimes forms a layer near the roof of a roadway, and it may move in the opposite direction to the ventilation if the current of air is weak.

Level A water course, sometimes used also for conveying coal.

Linesman A man who puts on survey lines, a surveyor's assistant.

Longwall A system of working coal originating in Shropshire where a number of men work along a coal-face. Roadways are maintained through the 'gob', and there are no pillars with this system.

Long-way A generic term covering longwall and various forms of bank work.

Low A light (eg candle).

Main road The main roadway driven from the pit-bottom through the workings.

Man-holes Refuge holes made along a roadway where men can shelter if, for example, a set of tubs becomes de-railed.

Marrow In Northumberland and Durham especially, a work mate; a member of a team paid on a common pay note. Men used to choose their own marrows.

Master-shifter In Northumberland and Durham, official in charge on a repairing shift.

Maul A wooden or iron mallet used to drive wedges into the coal-face and to hammer props up to the roof.

Meeting The place in the shaft where the ascending and descending 'bantles' of men passed each other and exchanged greetings when winding was slow.

Nick In bord-and-pillar, the hewer first undercut the coal, then cut a vertical 'nick' down each edge of the face of coal to be worked so as to loosen it.

Onsetter The man who operates shaft signals in the pit-bottom, and pushes the tubs on to the cage.

Opencast Working coal by quarrying.

Outbye See inbye.

Overcast An air crossing where one airway passes over another.

Overman An official inferior to the undermanager but superior to the deputy. In the North East the overman was originally in charge of a pit, usually responsible to a 'viewer' managing several pits.

Packs In hand-got longwall, pack walls used to be built to support roadways through the gob. In later practice packs were also built in strips in the area from which coal had been won allowing the roof to settle gradually. In modern practice, packs have largely been dispensed with and the roof settles quickly.

Panel working In bord-and-pillar, the workings are divided into districts (panels) enclosed within barriers of coal so as to restrict the spread of explosions and to facilitate the working of pillars with a minimum of convergence of the strata. 'Panel' is also loosely and misleadingly applied to longwall districts.

Pit An ambiguous expression. May mean (a) a single shaft (b) a pair of shafts (upcast and downcast) or (c) (nowadays) a colliery.

Props Pieces of wood or steel set vertically to support the roof. Horizontal bars were in later practice set up to the roof over the props.

Prove To confirm the existence or location of a seam of coal.

Putter A haulage hand employed on the subsidiary (not the main) roads underground.

Rag-and-chain An early type of pump operated manually or by animals or water-wheel.

Rake Rakes and forks (or screens) were used by hewers to separate the small coal from the large. The small coal was often left in the 'gob'.

Regulator A wooden partition with a shutter in an airway which restricts the flow of air inbye.

Return See intake.

Ride To travel up or down the shaft.

Rise See dip.

Rolley A wheeled carriage formerly used in Northumberland and Durham to carry several corves from the 'flat' to the pit-bottom.

Rook In the East Midlands, a capacity measure; stack of coal of a given size (say 2yd × 1yd × 1yd). Hence stacking coal was called 'rooking'.

Round coals Large coal from which the small has been separated.

Royalty The right to work minerals, usually vested in the freeholder.

Royalty rent (or royalties) Rent paid by the colliery proprietors for the right to work coal.

Scaffold A platform built (usually of wood) across a shaft rendering the lower part inaccessible.

Score In some districts hewers were paid so much per score of corves or tubs filled (not necessarily 20).

Scraper A wooden shaft with copper head for scraping coal dust out of a shot hole before firing.

Screen A fork for separating round coal from slack. (See also coal preparation plant.)

Set out A corf or tub of coal confiscated for containing less than a stipulated amount.

Shifters In Northumberland and Durham, men paid so much per shift, elsewhere called 'daywagemen' or 'datallers'. These were usually employed on repair work.

Sliding scale A method providing for wages to vary automatically with the selling price of coal.

Smart money Weekly payments to men off work through injury at work.

Snap In the Midlands, the mineworker's lunch.

Sough, suff or surf Drainage adit.

Staithes Coal wharves on the Tyne and Wear.

Stall-and-pillar Working by driving roadways ('stalls') into an area of coal, leaving pillars to support the roof.

Staple A shaft connecting one seam with another underground.

Steel mill A device for providing what was thought to be a safe light. A steel wheel is revolved against a flint from which sparks are emitted.

Stemming Clay or other material used to stem a shot hole before firing.

Stinkdamp (or sulphurdamp) Sulphuretted hydrogen.

Stopping A wall built to prevent air flowing beyond it.

Stover, stever or staver In the East Midlands, the senior member of a butty partnership taking charge of the surface arrangements.

Stythe Blackdamp.

Sump The wall at the bottom of a shaft from which water may be pumped.

Swalley, swelley or swilley A depression in the floor of the seam or roadway.

Tally stick Wooden stick inserted in coal tub to indicate which man or group produced the coal.

Ten A capacity measure equal to about 52 tons used as a unit in mineral rent agreements in some districts. The rent is called a tentale rent.

Thill The floor of the seam or roadway.

Thirl (or thurl) To cut through a wall of coal; to join two roadways by cutting through.

Tommy shop Shop owned by a colliery proprietor or butty where pay tickets were exchanged for goods under the Truck System.

Tram Originally, a wheeled carriage on which corves were placed on the main road. Now synonymous with 'tub'.

Tram plate Flanged or angle rail on which trams with plain wheels ran.

Trapper A boy employed on opening and shutting a ventilation door.

Truck system Payment of wages in kind.

Tub A small wagon used for conveying coal.

Tubbing Shaft lining of wood or iron to hold back the water.

Undercast An air crossing where one airway passes under another.

Underviewer The old term for undermanager; the official in charge of a mine in the absence of a viewer.

Upcast A shaft carrying an ascending air current. The top of an upcast shaft is always boxed in to prevent air from entering from the surface.

Viewer The old name for manager.

Wash-out Where a coal seam is intersected by the silt of a prehistoric river bed.

Waste Old workings in bord-and-pillar; the 'gob' in longwall.

Wasteman In bord-and-pillar, a man working in the waste on maintaining airways.

Wayleave rent Rent paid for the right to transport coal across someone else's land, whether surface or underground.

Whim-gin A horse-driven winding device.

Whimsey Term applied first to windlasses, then to whim-gins, but by early nineteenth century usually meaning an atmospheric winding engine. In South Yorkshire atmospheric pumping engines were also called whimseys.

Whin A hard stone sometimes encountered in sinking.

Whole 'Working in the whole' means working virgin coal by the bord-and-pillar method; as distinct from 'working in the broken' where only the pillars are left to be got.

Winning Opening out a mine so that it is ready to be used.

Working The art of producing coal.

Notes

These notes have been compiled for use in conjunction with the bibliography. References are given here in an abbreviated form, giving the author only or (where there is no author or more than one work by an individual author has been referred to) an abbreviated title. The full entries will be found in the list of secondary sources, primary sources (printed) or primary sources (manuscript) in the bibliography following this section.

INTRODUCTION

1   Swinnerton, p131; see also Francis, W., *Coal, Its Formation and Composition* (1954), *passim*
2   Lignite is found near Bovey Tracey and on the Sussex Weald, but there is none in any of Britain's coalfields. Martin, pp76–7
3   Nef, **I**, pp1–2; Taylor, T.J., pp3–6
4   Griffin, A.R., *Coalmining*, pp3–8; also 'Bell Pits', pp392–7
5   The Kentish coal measures lie at considerable depth, and are completely concealed beneath later formations. There were, therefore, no coalmines in Kent before the present century
6   Harrison, William, *Description of England* (1587), quoted in Wilson, J.D., *Life in Shakespeare's England* (Harmondsworth, 1944), p269
7   Gray, pp90–1
8   Evelyn, J., cited in Taylor, T.J., p9; Nef, **I**, p20
9   Gray, p91
10   *Ibid.*, pp84, 86
11   Smith, E., pp24–5
12   Nef, **I**, p20; Mitchell and Deane, p112. In converting from chaldrons to tons, I have accepted Mitchell and Deane's estimate of the London chaldron, ie $25\frac{1}{2}$cwt. Other authorities give it a slightly higher value, see Nef, **II**, pp367–8 and Bailey and Culley, p7
13   Ministry of Fuel and power, *Statistical Digest*, Cmd 6920 (1946), p8 (Table 1)

CHAPTER 1: THE ORGANISATION OF THE INDUSTRY

1   *VCH Notts*, **I**, pp270, 286–7
2   Nef, **I**, pp307–9; Gruffydd, K.L., 'Coalmining in Flintshire in the Sixteenth Century', *Buckley* (Magazine of the Buckley Society), No 2 (1972)
3   Nef, **I**, pp310–2; Coal Commission File DY 1240 (Derby C/RO 303/71); *RC on Mining Royalties*, Fourth Report, p32
4   Nef, **I**, p268; *RC on Mining Royalties*, Second Report, p116; *Sankey Report*, **III**, pp150–3
5   *RC on Mining Royalties*, Second Report, pp416–21
6   Nef, **I**, p315; *VCH Derbys*, **II**, p352; Galloway, **I**, p29; Dodd, p17
7   Smith, R.S., 'Willoughbys', pp77–80

8   Galloway, **I**, p75
9   *Ibid*, pp44–5
10 Nef, **I**, p320; Galloway, **I**, pp61, 70 (The number of men was often given as so many 'pickaxes'); *VCH Derbys*, **II**, p350; *VCH Notts*, **II**, p325; Galloway, **I**, p39
11 Nef, **I**, p321
12 *Ibid*, pp136–7; Marx, K., *Capital* (1867); *RC on Mining Royalties*, Fourth Report, p43; Moss, K.N., 'Mining Leases', in Moss, pp321–33; *VCH Notts*, **II**, p325
13 Smith, Adam, *Wealth of Nations* (1776), Everyman's Edition, **I**, p153
14 Nef, **I**, p326–7; *Sankey Report*, III, p16 (Appendix 10) and **II**, p690 (Evidence of T.H. Bailey)
15 Down and Warrington; Lease: Edge and others to Barber Walker, 1738, Notts CRO, DDE 14/45
16 Nef, **I**, pp150–6 and **II**, pp51–3; Galloway, **I**, pp95–8; Taylor, T.J., pp20–1, 63–76
17 Nef, **II**, pp58–65; Gray, pp86–7
18 *Ibid*, p86 (The £20,000 is an exaggeration. The actual amount was £6,000–£7,000)
19 Smith, R.S., 'Huntingdon Beaumont'; Nef, **II**, pp11, 150; Griffin, A.R., *Coalmining*, p14
20 Griffin, A.R., *East Midlands*, pp22–4
21 Griffin, C.P., *Leics. & S Derbys Coalfield*, pp238–40; Langton, J. 'Coal Output in South-West Lancashire' 1590–1799', *Econ Hist Rev*, **XXV**, (1972), p53
24 Galloway, **I**, pp53, 55, 75; Sitwell Mss
25 Sir Percival Willoughby to Huntingdon Beaumont, probably dated January 1602. Middleton Mss, cited in Smith, R S , 'Huntingdon Beaumont', pp133–4
26 Historical Manuscripts Commission, pp169–70; Griffin, C.P., *Leics & S Derbys Coalfield*, p41; Ashton and Sykes consider that the butty system developed from teams of equals working pits on contract in the course of the eighteenth century. A much earlier origin can be deduced from the Wollaton and Coleorton cases
27 J.C., pp31, 38, 42, 50
28 Smith, E., *passim*, especially pp31–2, 46
29 Indenture bond dated 8 October 1788, between John Nesham Esq and 105 workmen (Lound Hall Mining Museum)
30 Greenwell, pp3, 57, 60
31 Briggs, H., 'A Brief History of Mine Surveying', in Moss, p228
32 Galloway, **I**, pp260–1
33 There is a particularly useful section on this in Pollard, pp127–51 (Pelican edn); Galloway, **I**, pp308–9
34 Chaloner
35 The apprenticeship was almost certainly an informal one. Stephenson was at this time engineer at Killingworth Colliery (North of England Institute, p7)
36 Galloway, **I**, pp345–51
37 Duckham, *Scottish Coal*, pp113–40
38 *Mines Inspector's Report for the Midlands, 1886*, cited in Griffin, C.P., *Leics & S Derbys Coalfield*, pp300–1

39 *Mines Inspectors' Reports for 1869* (Mr Evans's Report), pp56–7
40 Skegby Colliery Accounts (Sutton-in-Ashfield Public Library); Pollard, p246, refers to this system as 'double-entry'. It only became double-entry if the entries were posted to another account (eg the Household Account). There is no indication of this at Skegby
41 Boot, also *Colliery Guardian* 20 August 1875 and 28 February 1876
42 Griffin, A.R., *East Midlands*, p31; *CEC,* First Report, App Pts I and II
43 *RC on Mines,* First Report, Cd 3548 (1907), p69 and Appendix XIV, pp404–5; The derivation of 'butty' is unclear. The Erewash term 'stever' ('stover' or 'staver') is clearly derived from to stave, ie 'make firm by compression' *(Concise Oxford Dictionary)* which probably has the same root as stevedore
44 Mee, *Aristocratic Enterprise,* pp94–108
45 *Ibid,* p114; Galloway, **II,** pp249, 251
46 Chambers, *Modern Nottingham;* Mee, *Aristocratic Enterprise,* pp139–42
47 *CEC,* App Pt I, p111
48 Griffin, C.P., *Leics & S Derbys Coalfield,* pp226–41; Beaumont, p9
49 Griffin, A.R., 'Thomas North'
50 Griffin, C.P., *Leics & S Derbys Coalfield,* 305–14; Griffin, A.R., *East Midlands,* p32
51 Lee, C., 'Introduction' to 2nd edn of Curr, pp1–3
52 Raybould (suggests that Beaumont's innovations had a long-lasting beneficial effect, but this seems unlikely. Indeed, while the facts given in this article are interesting, the argument is, in general, unconvincing)
53 North of England Institute, p7
54 *Ibid,* p9; Down and Warrington, p163
55 Denby Letter Book; 'The Derbyshire Coalfield, 15 Denby Colliery', *Colliery Guardian,* 2 December 1892 (I am indebted to Mr T.J. Judge for drawing this to my attention)
56 Goodchild, J., 'First Mines Inspector' pp15–17
57 Morton, C., (HMIM for the Midlands), *Report for 1851,* p2
58 North of England Institute, p11
59 Eg Griffin, A.R., *East Midlands,* p51

CHAPTER 2: CAPITAL

1 Galloway, **I,** pp52, 102
1 Historical Manuscripts Commission, p149 (In *Mining in the East Midlands,* the cost of this sough was given incorrectly as £20,000)
3 Smith, E., p24
4 Griffin, C.P., *Leics & S Derbys Coalfield,* pp5, 762; *Middleton Mss,* Huntingdon Beaumont to Percival Willoughby, in Smith, R.S., 'Huntingdon Beaumont', p138; Galloway, **I,** pp120, 176; J.C., p23
5 Nef, **I,** 54; Edge Mss DDE 46/67; Smith, R.S., 'Huntingdon Beaumont', pp120, 146
6 Edge Mss, DDE 5/21; *VCH Notts,* **II,** p326; Historical Manuscripts Commission, pp148-9
7 PRO E134/5W & M/Est 17, cited in Griffin & Griffin; Griffin, A.R., *Coalmining,* pp85–6

8   *RC on Mining Royalties,* Third Report 1891, p351; (Evidence of Thomas Chambers) also CEC, App Pt II, p338

9   Griffin, C.P., *Leics & S Derbys Coalfield,* pp7–8

10 Smith, E., p58. The use of the Scottish Quarter Day (instead of Lady Day) and also of the word 'chalder' (the Scottish form of chaldron) in Houghton Colliery reports suggests a connection with Scotland which is not apparent. For examples of long expensive soughs in this period see *VCH Derbys,* **II,** p354, Griffin, A.R., *East Midlands,* p5; Atkinson, p23; *Nef,* **I,** p356

11 Duckham, *Scottish Coal;* Galloway, **II,** p362; Bald, pp18–19

12 Griffin, C.P., *Leics & S Derbys Coalfield,* p23; Duckham, *Scottish Coal,* pp86–7, 147; Farey, **I,** p338; Galloway, **I,** pp265–6

13 Duckham, *Scottish Coal,* pp85–7; Goodchild, J., 'Steam Power', pp9–14; Nixon, pp17–22; Farey, *Agriculture of Derbyshire,* **I,** pp337–9, *Steam Engine,* p233; Bailey and Culley, p10; Ross, M. in Hair, pp16–17, 27, 44

14 *Ibid,* pp23, 29; Bailey and Culley, p11; Griffin, A.R., *East Midlands,* p63. See also chapter 5

15 *House of Lords Committee on the State of the Coal Trade* (1829), pp28–57

16 Ross, M. in Hair, p6; *SC on the State of the Coal Trade,* p266; *Midland Mining Commission,* pp cv–cviii. (For 37 collieries estimates of 'capital expended', totalling £2,475,000 were given. I have estimated the capitals of the 33 other collieries on the list by comparing their physical factors – depth, number of employees, horsepower employed, and output – with those of the 37 to reach the total of £4,336,000).

17 Holland, p189

18 Duckham, *Scottish Coal* pp147, 149–50, 182–3

19 Nef, I, pp19–20; also *Ibid,* **II,** Appendix B

20 The assumptions made are, in fact, dubious. If anything, the mines of Northumberland and Durham probably expanded comparatively more rapidly than those elsewhere between 1790 and 1829; the great expansion in inland coalfields followed the building of railways. On the other hand, generally the deeper the colliery, the more capital is required to produce a given output. Put another way, inputs of capital are subject to diminishing returns. Nevertheless, some such assumptions have to be made to arrive at a conclusion. An alternative method would be to list all collieries categorising them according to the level of technology, and attaching an average value to each category. Dr S.D. Chapman has done this for textile mills. At the time of writing the author has in hand an article in which this question is examined in detail. No allowance has been made in the estimates for stocks of coal or materials, hence the use of the expression 'fixed capital'.

21 *Mining Journal,* XX, p314, cited in Galloway, **II,** pp369–70

22 Griffin, A.R., *East Midlands,* pp22–4, 33; 'An Account of all the coals sold at sundry collieries in the Counties of Nottingham and Derby' (Ms), Notts CRO; Ashton and Sykes, p3

23 A 'stack load' at Wollaton and Strelley in the eighteenth century was 2½ tons. The Leicestershire load was probably the same. In 1727, Wilkins was selling coal at the pit-head at 8s 6d a load (equivalent to 4s 3d a ton if the load was 2 tons; and 3s 5d if it was 2½ tons).

24 Griffin, C.P., *Leics & S Derbys Coalfield,* pp12–21

25   Smith, R.S., 'Willoughbys', also 'Huntingdon Beaumont'; Edge Mss, DDE
     42/10. 11
26   Griffin, C.P., *Leics & S Derbys Coalfield*, pp4–6
27   *Ibid*, pp12–21
28   Duckham, *Scottish Coal*, pp147–54
29   *Ibid*, pp14, 174–99
30   Defoe, cited in Nef, **II**, p7; Galloway, **I**, p349–53
31   Nef, **II**, pp33–5; Smith, R.S., 'Huntingdon Beaumont', pp124–5
32   Galloway, **I**, pp93–5; Nef, **II**, pp30–43
33   *Ibid*, **II**, p61; Raistrick, A., *Quakers in Science and Industry* (Newton Abbot,
     1968), pp77–80ff; *Midland Mining Commission*, pp cv–cviii; Telford, S.J.,
     'Early Industrial Developments on the Northumberland Coast between Seaton
     Sluice and Cullercoats', *Industrial Archaeology* (August 1974), pp181–9; Dodd,
     pp306–13
34   Griffin, A.R., *East Midlands*, pp28, 107
35   Beaumont v Boultbee, 1797–1800, C12/225/1 (PRO), cited in Griffin, C.P.,
     Leics & S Derbys Coalfield, p25; also *ibid*, pp202–5
36   Griffin, A.R., 'Thomas North – Mining Entrepreneur Extraordinary',
     Transactions of the Thoroton Society (1972), pp53–73
37   Griffin, A.R., *East Midlands*, pp29–31, 38–43; Nef, **I**, pp426–7; Galloway, **I**,
     pp171–2; Dodd, pp366–7
38   Raybould, pp539–43. See also *Sankey Report*, Q14808, (evidence of J. Tryon)
39   *RC on Mines*, First Report, Cd 3548 (1907), 404; Taylor, A.J., 'The Sub-
     contract System in the British Coal Industry', *Studies in the Industrial
     Revolution* (1906), p217; *Midland Mining Commission*, p civ
40   Griffin, A.R., 'Thomas North', pp59, 67
41   Gosden, p167
42   Nef, **II**, p71
43   Mee, *Aristocratic Enterprise*, pp194–204; (see also *The Earls Fitzwilliam*,
     pp180–5)
44   Eg Coleorton with profits of £188 in 1576/7 and £248 in 1577/8, Middleton
     Mss, Mi Ac pp126–7
45   Middleton Mss; Miller-Mundy Mss, No 179 (Derby Borough Library – I am
     indebted to Mr P.S. Stevenson and Mr M. Burrows for this source)
46   *Sankey Report*, QQ 15846–51 (Evidence of John, Marquis of Bute)
47   Griffin, C.P., *Leics & S Derbys Coalfield*, pp27–8, 33, 205–7
48   Mottram and Coote, pp45–7, 77, 91, 116–20; Riden, p9
49   Griffin, A.R., 'Thomas North', pp53–73. In this article, the capacity of North's
     collieries is given as about 200,000 tons a year, but this underestimates his
     landsales which were proportionately much higher than his competitors'
50   Griffin, A.R., 'Boom Colliery'
51   Ross, M. in Hair, p43; Galloway, **I**, pp449–50, 472
52   One of them, Bobbie Shaftoe, was immortalised in a sea-shanty
53   Ross, M. in Hair, pp50–1
54   Griffin, A.R., *East Midlands*, p107; *Sankey Report*, **I**, p317
55   *Sankey Report*, QQ 23769–74 (evidence of Joseph Shaw) and Appendix 71, also
     QQ 19490, 19504–5 (evidence of T.J. Callaghan)
56   *Ibid*, Q 19671 (evidence of Lord Gainford) cf Q 21370 (evidence of W.

Thorneycroft)

57   *Ibid*, Q 19671; Labour Research Dept, *The Coal Crisis : Facts from the Samuel Commission* (1925–6), pp33, 35

58   *VCH Notts*, **II**, p329; *Annual Reports and Accounts of the Shireoaks Colliery Co Ltd* (1865–6); also Beaumont, p8. The sinking cost at Southgate (about £6 a yard) may be compared with the cost of sinking a 14ft diameter shaft at Moira (Leicestershire) in 1865 of £4 10s a yard plus £2 a yard for sinking through hard stone

59   *Sankey Report*, QQ 24–7 (evidence of A.L. Dickinson), also Vol **III** App 1, p3 and App 5, p7

60   *Ibid*, QQ 838–921 (evidence of J.C. Stamp) also Vol **III**, App 78, p230

61   Smith, D.L., *The Dalmellington Iron Company* (Newton Abbot, 1967), *passim*

62   Martin, E.A., *A Piece of Coal* (1896), pp118–48, 164–175; Lander, C.H., 'History of Coal Carbonisation', in Moss, pp257ff; Chaloner, W.H. and Musson, A.E., *Industry and Technology* (1963), Plate 41; Griffin, A.R., *East Midlands*, pp267–8; Levinstein, I., 'Observations and Suggestions on the Present Position of the British Chemical Industries, with special Reference to Coal Tar Derivatives' (1886), cited in Court, pp141–3

CHAPTER 3: LABOUR

1   Duckham, 'Serfdom', pp181–2

2   *Ibid*, p185, also diaries of the Earls of Wemyss

3   Duckham, 'Serfdom', pp188–9; Nef, **II**, p190

4   Duckham, 'Serfdom', pp193–6

5   Smith, E., pp9, 46

6   Galloway, **I**, pp269–70, 440–1, also Minutes of meeting of Tyneside Colliery proprietors dated 10 September 1805 *(Coals From Newcastle*, Doc 12)

7   Galloway, **I**, pp464–5, **II**, pp167–79; *SC on the State of the Coal Trade*, p274 (evidence of John Buddle); *An Appeal to the Public from the Pitmen of the Tyne & Wear* (Newcastle, 1832), pp4–5. (The owners were concerned about legal actions taken by the union's advocate W. Prowting Roberts, against oppressive enforcement of terms of the bonds. See W. Mitchell, *The Question Answered* (Bishopwearmouth, 1844) p18)

8   Galloway, **I**, pp167–9; Griffin, A.R., *East Midlands*, p160

9   Nef, **II**, pp151–3

10   Griffin, A.R., *East Midlands*, pp44–5.

11   Nef, **II**, pp135–6; Griffin, A.R., *East Midlands*, p35; Smith, E., pp24–5

12   Nef, **II**, pp138–40; Griffin, A.R., *East Midlands*, p35; *RC on Coal Supply* (1871), **III**, p32; Griffin, A.R., 'Industrial Archaeology', also *East Midlands*, pp16–17; Langton, J., 'Coal Output in South-West Lancashire', *Econ Hist Rev*, **XXV** (1972), p34

13   General Return of Men employed on the Tyne, 1815 *(Coals from Newcastle*, p21); *Lords Committee on the State of the Coal Trade* (1829), p54; *Midland Mining Commission*, pp cvii–cviii; Ross, M. in Hair, p6; Galloway, **II**, p178; Taylor, T.J., *Archaeology*, p58

14   Galloway, **II**, pp369–70; Mitchell and Deane, *Historical Statistics*, pp115, 118

15   *Mines Inspectors' Reports*

16   Lawson, J., *A Man's Life* (1932), pp56–8

17 Nef, **II**, p182; Middleton Mss; Galloway, **I**, p102; *VCH Notts*, **II**, p327
18 *VCH Derbys*, **II**, p355; Nef, **I**, p415 and **II**, pp190–6; Duckham, 'Serfdom', p189
19 Griffin, C.P., *Leics & S Derbys Coalfield*, pp81, 328; Nef, **II**, p182; J.C., p36; Ashton, 'Coalminers' pp314–7 Newbottle Miners' Bond; Mee, 'The Earls Fitzwilliam', p311
20 *CEC*, App Pt 2, *passim*; *VCH Derbys*, **II**, p355; Riden, *Butterley Company 1790–1830*, p29; Griffin, C.P., *Leics & S Derbys Coalfield*, p329; *East Midlands*, pp36–7; Machen, p56
21 Galloway, **I**, p468 and 2, p177; Greenwell, pp47, 75; *Observations on the Laws Relating to the Colliers in Scotland, etc* (Glasgow, 1825), pp49–50
22 Nef, **II**, pp195–6; Griffin, A.R., *East Midlands*, pp35–7; Mee, The Earls Fitzwilliam, p298; Chambers, *Workshop*, p22
23 Ashton, *Industrial Revolution*, p117; Chambers, *Workshop*, p221; Griffin, A.R., *East Midlands*, pp36–7, 57, 115, 126–7; Derbys CRO, Ref DRO D535
24 White, pp167, 175–6; Griffin, C.P., *Leics & S Derbys Coalfield*, pp74–8
25 Baines, E., p95; *CEC*, App Pt I, pp112–7; Griffin, A.R., *East Midlands*, pp112–4, 129, 184; Rowe, pp72, 82; Slaven, p234; Aitken, J., 'Observations at Hawarden, 1861'; Buckley, No 2 (1971)
26 *CEC*, App Pt II, pp303, 313; Mee, The Earls Fitzwilliam, p268; Straker, W., *Acts of Parliament Affecting Mines* (Newcastle, 1935), pp292–3; Gosden, pp32, 62, 110–1; Wilson and Levy, I, *passim*
27 Griffin, A.R., *East Midlands*, pp37–8; White, p170; *CEC*, *passim*; Greenwell, pp75, 88; Parkinson, pp10, 22
28 Nef, **II**, pp167–8; Duckham, *Scottish Coal*, pp95–102, 279–83; Griffin, A.R., *East Midlands*, pp42, 172; Galloway, **I**, pp91, 232–4, 305, 354, 503, 2, pp149–51, *CEC*, *passim*; *VCH Derbys*, **II**, p351; Engels, p276; Hallam, pp18, 66
29 Griffin, A.R., *East Midlands*, pp79–80, 259, 272, 309–10, 316; Griffin, A.R., 'Industrial Relations'; Mitchell, W., pp14–15
30 Griffin, A.R., *East Midlands*, p39; Griffin, A.R., 'Contract Rules'; Nef, **II**, pp155–6
31 White, pp171–2; Griffin, A.R., *East Midlands*, pp30, 39–41
32 Galloway, pp167–8; *Midland Mining Commission*, pp35–7; Griffin, A.R., *East Midlands*, pp40–1; There is evidence for the survivial of boon work at Darley and Ashover in Derbyshire in the seventeenth century where certain tenant farmers were bound to carry a load of coal yearly for the lords of the manor (*VCH Derbys*, **II**, p352)
33 Griffin, A.R., *East Midlands*, p41; Nef, **II**, p188; Mee, The Earls Fitzwilliam, p320; Griffin, C.P., *Leics & S Derbys Coalfield*, pp106–7
34 Ashton and Sykes, p130; Griffin, C.P., *Leics & S Derbys Coalfield*, pp101–8
35 Duckham, *Scottish Coal*, pp305–7; Webb, pp44, 89–90
36 Arnot, **I**, pp36–8; Webb, p124; Griffin, A.R., *East Midlands*, pp74–5; Galloway, **I**, pp465–70; Mitchell, W., p21; Scrutator, *An Impartial Enquiry into the . . . causes of dispute between the Coal owners . . . and their late Pitmen* (Houghton-le-Spring, 1832) pp6–8; Slaven p219
37 Griffin, C.P., 'Chartism: a critical note' and 'a reply'
38 Galloway, **II**, pp173–9; Griffin, A.R., *East Midlands*, pp71–5; Arnot, pp41–3;

Webb, pp181–6, 299–300

39 Machen, p287; Griffin, A.R., 'Thomas North', p61; Stuart Smith, p159

40 Griffin, A.R., *Miners* (1956), pp20, 33; Webb, pp304–6; Griffin, A.R., *East Midlands*, pp76–88

41 Arnot, p61, *Colliery Guardian*, Jan 1 & 29, Feb 5 & 26, April 2, 9, 16 & 30, June 11, and Aug 13 1875 also Oct 13 1876

42 *Ibid*, pp91–120

43 Griffin, A.R., *Miners* (1956) pp145–6, and *East Midlands*, pp149–50; Ashton T., p197 (Speech of Mr. A. M. Chambers)

44 Ashton, T., p236; Griffin, A.R., *East Midlands*, pp152–9; Williams, 'Labour' and 'Miners'; Porter, J.H., 'Wages Bargaining under Conciliation Agreements', *Econ Hist Rev*, **XXIII** (1970), p474

45 Rowe, pp126–7; Griffin, A.R., *East Midlands*, pp182–4

46 Griffin, A.R., 'Miners' Lockout', pp64–5

47 Jevons, p359. The rates quoted are basis rates to which current percentages need to be added

48 Arnot, pp192–8, 369–70

49 *Ibid*, pp287–8, 298

50 Griffin, A.R., *East Midlands*, pp205–7; Phelps-Brown, pp318–20, 355; Unofficial Reform Committee, *passim*; Jevons, pp520–55; Griffin, A.R., 'Miners' Lockout', p63

51 Rowe, pp102–10

52 Griffin, A.R., *East Midlands*, pp202–22

53 The author has written extensively on this topic elsewhere, see, eg *East Midlands*, pp217–315; *Miners of Notts* (1962), *passim*. A new account of the general strike and its antecedents will be found helpful: this is Renshaw, P., *The General Strike* (1975). See also Arnot, R. Page, *The Miners: Years of Struggle* (1953) 393–523

CHAPTER 4: TECHNOLOGY

1 Griffin, A.R., *Coalmining*, pp6–7, 47–50

2 *Ibid*, pp7–10, 50–2

3 *Ibid*, pp29–30, 53–6

4 *Ibid*, pp12–15, 41–3, 53–5; Lewis, pp104–7; Griffin, C.P., *Leics & S Derbys Coalfield*, p138

5 Griffin, A.R., *Coalmining*, pp15–18; also Curr; and Lewis, p317

6 Curr, pp2–3; Galloway, **I**, p323

7 Galloway, **I**, p332

8 Griffin, A.R., *Coalmining*, pp55–6; Taylor, A.J., 'Sub-Contract System', pp221–2; Farey, *Derbyshire*, **I**, pp341–51

9 Griffin, A.R., *Coalmining*, p26

10 *Ibid*, pp84–92, 94–103; Sheard and Hurst, p558; Duckham, *Scottish Coal*, pp79–80

11 Griffin, A.R., *Coalmining*, pp62–83

12 *Ibid*, pp30–40; Curr, p34; Galloway, **II**, pp326–7. About 1840, winding engines in the North East were from 20 to 50hp but there was an exceptionally large one of 80hp at Jarrow

13 Galloway, **I**, pp481–4 and **II**, pp33–4

14   *Ibid,* **I,** pp485–6. The earliest recorded use of gunpowder for coal-getting was at Hetton in 1813, but its general use came twenty years or more later

15   Galloway, **II,** pp241–8; Griffin, A.R., *Coalmining,* pp52–6; Anderson, D., 'Blundell's Collieries, Technical Developments, 1776-1966', *Trans Hist Soc of Lancs & Cheshire* (1957), pp123–5. Some stall and pillar work remained in all these districts, however

16   Galloway, **II,** pp233, 249–51; Taylor, T.J., p52

17   Griffin, A.R., 'Thomas North', pp64–6; also Stuart Smith, pp155–6

18   Morton, C., (HMIM for the Midlands), *Report for 1851,* pp6, 10

19   *Midland Mining Commission,* p69; Ashton, *Industrial Revolution,* p35; *RC on Mines,* First Report, Cmnd 3548 (1907), pp69–81

20   Griffin, A.R., *Coalmining,* pp53–61, 115; *Mines Inspectors' Reports,* 1912-1938; Ministry of Fuel and Power, *Statistical Digest*

21   Griffin, A.R., *Coalmining,* pp71–5

22   *Ibid,* pp116–22; Byres, T.J., 'Entrepreneurship in the Scottish Heavy Industries, 1870-1900', *Studies in Scottish Business History,* p251

23   Griffin, A.R., *Coalmining,* pp123–5

24   Iron props are sometimes heard of, usually at mines owned by ironmasters (eg Butterley), but they formed an inconsiderable proportion of the total

25   Griffin, A.R., *Coalmining,* pp127–8

26   See Duckham, *Pit Disasters*

27   Galloway, **I,** pp422–35. For a pro-Stephenson view, which uses only part of the available evidence, see Rolt, pp24–34; *House of Lords Committee on the State of the Coal Trade* (1829), p32

28   Where there is between 5 and 15 per cent of firedamp (methane) in the air, the mixture is explosive. With a higher percentage, the firedamp will burn but not explode

29   Morton, C., (HMIM for the Midlands), *Report for 1851,* pp8–9

30   Acetylene (carbide) lamps, naked flame lamps and candles were still permitted at mines not thought to be affected by firedamp

31   On underground lighting, sources consulted include Galloway, *passim, Practical Coalmining* (pub Gresham in numerous editions) and Mason, **I,** pp231–61

32   Stuart Smith, pp158–9; Dron, R.W., 'Lighting of Mines', in Moss, pp167–8; Hardwick, F.W. and O'Shea, L.T., 'Notes on the History of the Safety Lamp', *Trans Inst Min Engrs* (1915–16), p654

33   Haldane, J.S., 'Health and Safety in British Coal Mines', in Moss, pp272–5; NCB *Coal Dust* (1960 and 1961) and *Stone Dust*

34   McCutcheon

35   This ambiguous use of the term chokedamp has misled many unwary historians. For example, the compilers of *Coals from Newcastle,* which is otherwise excellent, regard the terms 'chokedamp' and 'afterdamp' as synonymous. Again R.A. Buchanan, in his disappointing account of coalmining, suggests that canaries taken into the mine were 'the traditional remedy' for chokedamp. In fact, canaries are taken underground after explosions because they are more sensitive to carbon monoxide (afterdamp, not chokedamp) than human beings and therefore give an early indication of its presence. In any case, it is no more sensible to regard canaries as a remedy for gas than it would be to regard a

clinical thermometer as a remedy for influenza. (See Buchanan, p77)

36 Bryan, Sir Andrew, *Report on the Accident at Creswell Colliery, Derbyshire*, Cmd 8574 (1952)

37 *Mines Inspectors' Reports for 1867*, pp5, 11; *Colliery Year Book* (1951 edn), pp590–1; Ministry of Fuel and Power, *Statistical Digest* (1951), Table 52, p75. (NB There is some slight inconsistency between the figures in the *Colliery Year Book* and the *Statistical Digest*;) NCB *Reports and Accounts*

38 Morton, C., *Report for 1851*, pp7, 11, 12

39 And also the best industrial relations. See Griffin, A.R., 'Consultation', pp29–30, 46–7

40 Pay Board, p13

CHAPTER 5: THE COAL TRADE

1 Nef, **I**, pp84, 92. Nef's estimate for Scottish coal exports has recently been questioned by Smout and Duckham (Duckham, *Scottish Coal*, p226)

2 *Ibid*, **I**, pp53, 81–3

3 Middleton Mss, p175; (The Wollaton Rook at this date measured $2\frac{1}{4}$yd high by 1yd square, close stacked, equivalent to perhaps $1\frac{1}{2}$ tons); Nef, **I**, pp96, 110; Griffin, A.R., *East Midlands*, p62

4 Nef, **I**, 101–2; Griffin, A.R., *East Midlands*, p63; Duckham, *Scottish Coal*, p205

5 *VCH Derbys*, **II**, p352; Griffin, A.R., *East Midlands*, p62; Coalmining, pp12–15; Taylor, T.J., p34; Rees; Duckham, *Scottish Coal*, p210; Lewis, pp86–109, 125–36, 225, 232–40. (Rees gives 1738 as the date of the Whitehaven Wooden Wagonways)

6 Griffin, A.R., *Coalmining*, p23; Atkinson

7 The Newcastle chaldron was equivalent to about a ton in the fifteenth century, but increased gradually in size (no doubt to lighten the tax burden) to 53cwt in 1694. The London chaldron was approximately half this. (See Nef, **II**, pp367–70; Taylor, T.J., pp20–4

8 Forster, p14

9 Taylor, T.J., p44; Forster, pp7–10

10 Defoe, *passim*; Duckham, *Scottish Coal*, pp205–8; Griffin, A.R., 'Thomas North', p53; Griffin, C.P., *Leics & S Derbys Coalfield*, pp792–801; Riden, *Butterley Company 1790–1830*; p28; Rees, Article on Railways

11 Taylor, T.J., p33; The Newcastle chaldron was at one time 42cwt: Duckham, *Scottish Coal*, p207; Chambers, *Vale of Trent*, p13

12 Griffin, A.R., *Coalmining*, pp21–2; Hassall and Trickett; Griffin, A.R., *East Midlands*, p63; Griffin, C.P., *Leics & S Derbys Coalfield*, p784

13 Duckham, *Scottish Coal*, pp217–22, 233

14 Griffin, C.P., *Leics & S Derbys Coalfield*, pp804–5; Griffin, A.R., *East Midlands*, p63; Lewis, pp231, 261

15 Griffin, A.R., *Coalmining*, pp14–19; Ripley, *passim*; Hopkinson, p28; Riden, 'Butterley Company', pp33, 48

16 Goodchild, *South Yorkshire*, p2; Lewis, p148

17 *Repertory of Arts and Manufactures for 1800*, cited in Griffin, A.R., *East Midlands*, p63; also Rees, Articles on Canals. An estimate for a Scottish line about 1810 was £837 10s per mile: Duckham, *Scottish Coal*, pp209–17

18 Griffin, A.R., *Coalmining*, pp18–20

19 Galloway, **I**, pp452–3
20 *Ibid*, **I**, pp455–6; Hair, p51; Galloway, R.L., *The Steam Engine and its Inventors* (1881), pp229–31
21 Griffin and Griffin, pp103–4; Griffin, C.P., *Leics & S Derbys Coalfield*, pp812–24. (The function of engineer for the Leicester to Swannington Railway was undertaken by George Stephenson's son Robert)
22 The Moira Collieries, for which the Woodhouses were the consultant engineers, were also fairly modern, but although in the county of Leicester are usually regarded as South Derbyshire mines (Griffin and Griffin, p120)
23 See eg Griffin, A.R., *East Midlands*, pp97–108, 160–7
24 Nail, p3
25 *Ibid*, pp1–6; Taylor, T.J., pp17–18; Nef, *passim*. For a full account of the system of metage see the *SC on the State of the Coal Trade* (1830), pp3–18
26 Farey, *Derbyshire*, **I**, p186; Nail, pp2–3
27 Cited in Beaumont, p4
28 Butterley Sales & Output Book, cited in Griffin, A.R., *East Midlands*, pp98–9
29 Griffin, A.R., *East Midlands*, p99
30 Mitchell and Deane, p113
31 *Ibid*, p121; *Colliery Year Book* (1951), p587
32 Griffin, A.R., 'Thomas North', p65, and *East Midlands*; Griffin and Griffin, p102
33 Ministry of Fuel and Power, *Statistical Digest* (1951), p93
34 Gray, pp80, 90–1; Nef, **I**, pp190–215
35 Gale, pp31–4
36 Lauder, C.H., 'The History of Coal Carbonisation', in Moss, pp253–65
37 Griffin, A.R., 'Contract Rules'; also 'Checkweighing', and *East Midlands*, pp79–80, 92
38 Ross, M. in Hair, p8; Griffin, A.R., *Coalmining*, pp106–9
39 Miners' coal for the years 1869, 1887 and 1903 accounting for about 3 per cent, has been transferred from the 'General Industrial and Miscellaneous' column to the 'Domestic' column for the sake of consistency
40 *Colliery Year Book* (1951), pp532, 576. (There were some imports of coal in the years following World War II amounting to 0·7 million tons in 1947 and 0·1 millions in 1949)
41 *SC on the State of the Coal Trade*, p6; and Bland, Brown and Tawney, pp497–9
42 Ross, M. in Hair, p5
43 *RC on Coal*, Report 3, p12; cited in Jevons, p316
44 Taylor, A.J., 'Combination'; Smith, R., pp281–3; Galloway, **II**, p369; Griffin and Griffin, p95
45 For example, the Denaby Colliery Company, formed in 1868, was capitalised at £110,400. Sinking through heavily waterlogged ground took four years, the Barnsley bed being reached at 448 yards. See Ball, p10; also Wilcockson, W.H., *Sections of Strata of the Yorkshire Coalfield* (1950), pp121–2
46 Griffin and Griffin, *passim*
47 *Ibid*, pp98–9; Duckham, *Scottish Coal*, pp234–8
48 Raybould, pp533–4
49 Court
50 Griffin and Griffin, pp95–7, 101; Griffin, A.R., *East Midlands*; Nef, **II**, p115

51 Trade and Navigation Accounts, cited in *Colliery Year Book* (1951), p559. (I have taken the national average pit-head price to be 60 per cent of the FOB export price. In 1884, when the average export price was 9s 2d the average pit-head price was 5s 5d, whilst in 1895 when the average export price was 9s 3d, the average pit-head price was 6s. This relationship appears to be fairly constant for the late nineteenth century)

52 Information re Attorney General v The Great Northern Railway Company, cited in Griffin, A.R., 'Thomas North', pp68–9, 73; also Smith, R., p284

53 Ball, pp34–5

54 Griffin, A.R., *East Midlands*, pp98–9. This agreement brought to an end a rate war between the two companies which had lowered rates to London in the 1860s: see Griffin, C.P., *Leics & S Derbys Coalfield*, pp827–8

55 Williams, *Derbyshire*, p185

56 *VCH Notts*, **II**, p325

57 The elasticity of demand was low in summer, but high in winter. See Burnett, p138

58 Nef, **I**, pp397–8 and **II**, pp72, 78. (One woodmonger whose business still exists was William Cory)

59 Forster, pp22–3

60 J.C., pp50–2

61 Defoe, pp315–6. See also Defoe's account of the coal trade in *The Complete English Tradesman*, (cited in Bland, Brown & Tawney, pp491–2) and Smith, R., pp37, 78. At first, lightermen sometimes offered shippers prices higher than those current in the market so as to drive independent competitors out

62 Nef, **I**, p410. Watermen were given the monopoly for carrying passengers

63 Horne, pp10, 18; Smith, R., pp 41, 47, 66–79, 106–7, 147, 173

64 Nef, **II**, p98; Smith, R., pp39 *et seq*

65 The emphasis is Mayhew's. Mayhew p554; Smith, R., pp91, 232–75

66 Smith, R., p161

67 Shireoaks Colliery Co Ltd Reports; Smith, R., pp277–83; Heys, pp46–50

68 Griffin, A.R., *East Midlands*, pp67–8

69 *Sankey Report*, QQ 1223–60

70 *Ibid*, Q 9629. Also Redmayne, pp116–8. According to Smith, R., (p343), the Cory combine, established in 1896, controlled 70 per cent of the sea-borne imports and over a third of the total imports into London

71 *Sankey Report*, App 46 (Vol **III**, p73). See also *Colliery Year Book* (1951), pp564–72

72 Tawney, pp133–4

73 Burnett gives the retail price of a ton of best coal in London as 39s in 1815, 17s 3d in 1845 and 21s just before World War I. Since he believes that coal duties were abolished in 1835, and does not give his sources, his figures need treating with caution (Burnett, pp217–8)

74 Mayhew, pp261–2

CHAPTER 6: MINING COMMUNITIES

1 Nef, **I**, p19, and **II**, pp149–50

2 Leifchild, pp196–7

3 Duckham, *Scottish Coal*, p259

208

4   Holland, p292; Leifchild, p190
5   Jevons, p651
6   Liberal Party, pp123–39. In the 1860s, houses 200 years old were still occupied
     in parts of Scotland. They were one-roomed cottages, 12 feet by 15 feet (with
     perhaps a tiny loft), had earthen floors, and were built of crude masonry without
     foundations (Duckham, p257); there is little doubt that some of those reported
     on by Scott-James were identical with those in the 1860 report
7   Griffin, A.R., *Coalmining*, p132–6
8   Leifchild, pp191–3
9   Duckham, *Scottish Coal*, p259
10   Griffin, A.R., *East Midlands*, pp160ff
11   Griffin, A.R., 'Methodism', and 'Primitive Methodism'; Holland, pp293–6;
     Griffin, A.R., *East Midlands*, pp44–52
12   Tremenheere, S., *Report on the Mining Population in parts of Scotland and
     Yorkshire* (1845), p25, cited in Mee, The Earls Fitzwilliam, pp270–1
13   Mee, The Earls Fitzwilliam, pp273–5; *CEC* (Report of J.C. Symond, Sub-
     Commissioner for the West Riding)
14   Griffin, C.P., *Leics & S Derbys Coalfield*, pp506–11
15   Chambers, *Modern Nottingham*
16   MacFarlane; Ball, p17; Griffin, A.R., *Miners* (1962), 2, pp255–75. Denaby's
     history is curious. Although well to the east of the old Yorkshire mining area, it is
     not on the concealed coalfield. The Shafton coal was found in the Denaby shafts
     at a depth of just under 11 yards, and it is clear from an advertisement in the
     York Courant for 12 September 1769 that this coal was being worked, on quite a
     large scale for the period, a century before Denaby Main was opened. This
     advertisement offers for sale 'at Denaby, near Doncaster, one Cylinder, 55
     Inches in Diameter, with the necessary appertenances thereto belonging, Two
     sets of pumps, with the Wind-Bores and working Barrels of 20 Inches Bore, for
     40 Yards deep, with the Necessities thereto. Three Coal Gins, 11, 12 and 14 feet
     Diameter. Four Coal Waggons with Metal Wheels, a Parcel of Bore-Rods, and a
     large Quantity of both new and old Waggon Rail and Sleepers together with a
     Quantity of different sorts of Timber and Plank and several other Materials
     relative to a colliery'. Application was to be made to Aaron Walker, of
     Rotherham, Jonathan Smith of Ravenfield or William Lyster at Haugh Colliery,
     near Rotherham. (I am indebted to Dr M.J.T. Lewis for this reference).
     The technical reason for abandoning the old Denaby Colliery once the Shafton
     coal was substantially exhausted is simple: the next workable seam (of
     indifferent quality, too) was the Double Smuts lying at a depth of over 80 yards,
     while the main seam, the Barnsley Bed, was 368 yards deeper still. What is
     strange is that, in less than a century, the mining tradition should have
     disappeared so completely. See Wilcockson, p121; and Lewis, p133
17   Storm-Clark
18   Griffin, A.R., *East Midlands*, pp111–14; Griffin, C.P., *Leics & S Derbys
     Coalfield*, pp246–8
19   *Derbyshire Times*, 7 November 1901; *Colliery Guardian*, 6 April 1900; Griffin,
     C.P., *Leics & S Derbys Coalfield*, pp441–74
20   National Coal Board, *Creswell*
21   Griffin, A.R., *East Midlands*, pp168–70

22   Jevons, pp155–75

CHAPTER 7: THE COAL INDUSTRY IN DECLINE

1   Buxton; Griffin, A.R., 'Industrial Archaeology'; Ministry of Fuel and Power, *Statistical Digest* (1951), p23; *Samuel Report*, I, p265; *Buckmaster Court of Inquiry into the Wages Position of the Coalmining Industry* (1924), 5, p287
2   *Colliery Year Book* (1951), p565
3   *Reid Report*, p16; *Report of Secretary for Mines* (1934), pp14–15
4   Griffin, A.R., *East Midlands*, pp256–9; Neuman, pp161–5; *Rules of the CCCA* (Leeds, 1928); *Report of Secretary for Mines* (1934), pp132–3; Kirby, M.W., 'The Control of Competition in the British Coalmining Industry in the 1930's', *Econ Hist Rev*, 2nd ser, **XXVI** (1973), pp276–7
5   Foot, p5; *Reid Report*, p139
6   National Coal Board, *Report and Accounts* (1947), pp10, 40, 193–4
7   The NCB were not involved, and had no interest in, the allocation of the global sum. Mr C. Storm-Clarke retails an apocryphal story about a colliery company claiming for a 'dummy' colliery and implies that this would be of concern to the NCB ('The Miners, 1870-1970: A Test Case for Oral History')
8   National Coal Board, *Report and Accounts* (1947), pp90–1
9   Ezra, pp9–10
10 Wilberforce, pp4–5; also Griffin, A.R., 'Consultation', pp30–1
11 *Fuel Policy*, Cmnd 3438 (1967), pp17, 37
12 *Notts Miners* II, 290–303
13 See Robinson

POSTSCRIPT

1   Ezra; National Coal Board, *Report and Accounts* (1970–1), pp2, 5
2   *Ibid*, pp2, 6–7
3   *Coal Industry Bill, Explanatory and financial memorandum*
4   National Coal Board *Report and Accounts* (1973–4), pp5, 8, 29, 43
5   Ezra, p10; National Coal Board, *Plan for Coal*, p7
6   Chandler, p22 (A ton of oil is, on average, equal to 7·45 barrels)
7   Ezra, p8
8   National Coal Board, *et al*, pp2–4
9   Varley, pp3–5
10 Department of Energy, *Interim Report* pp10–11
11 1 ton equals 1.016 tonnes. In West Germany, the daily output per mine in 1969 was 6,606 tonnes against 1,990 tonnes in the UK
12 Chandler, p24; also *Financial Times*, 18 July 1974 (article by Adrian Hamilton)
13 Varley, p7; also *Guardian*, November 1976 (article by Dennis Johnson)
14 *Financial Times*, 30 October 1976 (article by Roy Hodson); also *Inbye*, Number 18, November 1976
15 *Talking about Coal*, Number 121, November 1976

Bibliography

PRIMARY SOURCES, MANUSCRIPT

Babbington Colliery Co Memorandum Book, 1839–59 (NCB South Notts Area)
Boot, John, Report Book of Inspections at Strelley (Notts CRO, DDE 28/30)
Booth, H.W., letters to A.R. Griffin
Brown, William, Letter Book (North of Eng Inst of Mining & Mech Engrs)
Butterley Company, Sales and Output Book (Derbys CRO)
Butterley Company, Statistical Survey 1856 (Derbys CRO)
Chatsworth Estate Plans (Mines Records Office, Eastwood)
Coal Commission File DY 1240 (Mines Record Office, Eastwood)
Coal Owners' Associations, documents concerning, (Notts CRO DD528/1–73)
Denby Drury-Lowe, Colliery Letter Book 1885–6 (Lound Hall Mining Museum)
Drury-Lowe Mss (University of Nottingham)
Dunn, Mathias, 'History of the Viewers' (North of Eng Inst of Mining & Mech
    Engrs)
Earls of Wemyss, Diaries (typed extracts, NCB Scottish Area)
Edge Mss (Notts CRO)
General Return of Men employed on the Tyne, 1815 (Coals from Newcastle, Doc 21)
Haslam, J. et al, 'A Logue Book of Miners and Mining Information', (Ilkeston Public
    Library)
MAD Scheme – Notts & Derbys Section, Minutes, 1939 (typescript in private
    possession)
Miller-Mundy Mss (Derby Central Library)
Middleton Mss (University of Nottingham)
Miners' Bond, Newbottle Burn Moor Colliery, 1788 (Lound Hall Mining Museum)
Minutes of Meeting of Tyneside Colliery Proprietors, 10 September 1805 (Coals
    from Newcastle, Doc 12)
Sitwell, Mss and Plans (Derbys CRO and Renishaw Hall)
Skegby Colliery Account Book 1847–8 (Sutton-in-Ashfield Library)
South Normanton Colliery Co Ltd, Reports and Accounts, 1892–1904 (Lound Hall
    Mining Museum)
Stanton Ironworks Co, Notice of Wage Reduction at Teversal, 8 March 1876
    (Lound Hall Mining Museum)
Wallsend Main Colliery Co, Barnsley, Report on the colliery workings and prospects
    by William Sutton, 25 February 1903 (Lound Hall Mining Museum)

PRIMARY SOURCES, PRINTED

Bryan, Sir Andrew, *Report on the Accident at Creswell Colliery, Derbyshire* (1952)
CEC: See *Children's Employment Commission*
Central Collieries' Commercial Association, Rules (Leeds, 1928)
Children's Employment Commission, Mines, 1842
Coal Industry Bill (Bill 40, 1972)
Coal Industry Commission Report (Sankey Report), Cmd 361 (1919)

Coal Mines Acts (various dates)
Coalmining : Report of the Technical Advisory Committee (Reid Report), Cmd 6610 (1945)
Department of Energy, *Coal Industry Examination, Interim Report* (June 1974)
Department of Energy, *Coal Industry Examination,* Final Report (December 1974)
DSIR, *The Concealed Coalfield of Yorkshire and Nottinghamshire* (1951 edn)
Historical Manuscripts Commission, *Report on the Mss of Lord Middleton,* Cmd 5567 (1911)
Home Office, *Lists of Coalmines* (various dates from 1862)
Information re Attorney General *v* The Great Northern Railway Company, 1859
Institution of Mining Engineers, *Proceedings* (various dates)
Midland Mining Commission, First Report (1843)
Mineral Statistics (various dates from 1854)
Mines Inspectors' Reports (various dates from 1851)
Miners' Federation of Great Britain, *Minute Books* (various dates from 1889)
Ministry of Fuel and Power, *Statistical Digest* (various dates)
National Association of Colliery Managers, *Proceedings* (various dates)
National Coal Board, *Annual Reports and Accounts* (various dates)
Newspaper Files: *The Derbyshire Times; Nottingham Review; Nottingham Express; Colliery Guardian*
Pay Board, *Special Report, Relative Pay of Mineworkers,* Cmd 5567 (1974)
RC on Coal (1870)
RC on Coal Supply (1871)
RC on Mining Royalties (1891)
RC on Mines, First Report, Cmd 3548 (1907)
RC on Coal Industry (Samuel Report), Cmd 2600 (1925)
Reid Report : See *Coalmining : Report of the Technical Advisory Committee*
Samuel Report : See *RC on Coal Industry*
Sankey Report : See *Coal Industry Commission Report*
SC on the State of the Coal Trade (1830)
SC on the Causes of the Present Dearness and Scarcity of Coal (1873)
Secretary for Mines Reports (various dates)
Shireoaks Colliery Co Ltd, Annual Reports and Accounts 1865–7
Shireoaks Colliery Co Ltd, Reports of C. Tylden Wright, dated 30 January 1867 and 17 July 1867, and of John Hedley, dated 3 August 1867, regarding the proposed development of a coalmine at Southgate
Wilberforce, *Report of Court of Inquiry into a dispute between the NCB and the NUM,* Cmd 4903 (1972)
Wilcockson, W.H., *Sections of Strata of the Coal Measures of Yorkshire* (1950 edn, Sheffield)

SECONDARY SOURCES

Arnot, R. Page, *The Miners* (1949)
Ashton, T., *Three Big Strikes in the Coal Industry* (Manchester, nd)
Ashton, T.S., *The Industrial Revolution* (1948); 'The Coalminers of the Eighteenth Century', *Economic History,* **I** (1928)
Ashton, T.S. and Sykes, J., *The Coal Industry of the Eighteenth Century* (Manchester, 1929)

Atkinson, F., *The Great Northern Coalfield* (Newcastle, 1966)

Bailey, J. and Culley, G., *General View of the Agriculture of Northumberland, Cumberland & Westmorland* (1805)

Baines, E., *The Woollen Manufacture of England* (1875, reprinted Newton Abbot, 1970)

Bald, R., *A General View of the Coal Trade of Scotland* (1812)

Ball, B., 'Denaby Main, A Coalmining Village from 1850 to the Denaby and Cadeby Collieries Dispute (Bag-Muck) 1902-3' (unpublished Diploma Thesis, Ruskin College, 1974)

Beaumont, P., *History of the Moira Collieries* (Derby, 1919)

Bland, A.E., Brown, P.A. and Tawney, R.H., *English Economic History – Select Documents* (1914)

Buchanan, R.A., *Industrial Archaeology in Britain* (1972)

Burnett, J., *A History of the Cost of Living* (Harmondsworth, 1969)

Buxton, N., 'Entrepreneurial Efficiency in the British Coal Industry between the Wars', *Econ Hist Rev*, 2nd ser, **XXIII**, No 3, 1970, pp476-97

Chaloner, W.H., 'James Brindley (1716-72) and his Remuneration as a Canal Engineer', *Trans Lancs & Chesh Antiquarian Soc*, Vols 75, 76, 1965-6

Chambers, J.D., *Modern Nottingham in the Making* (1948)
    Workshop of the World (1961)
    The Vale of Trent, 1670-1800 (*Econ Hist Rev* Supplement, **3,** nd)

Chandler, G., 'Energy, the international compulsions', *Coal and Energy Quarterly*, No 1, Summer 1974, pp20-6

Colliery Year Book and Coal trades Directory (Annual)

Court, W.H.B., *British Economic History 1870-1914* (Cambridge, 1965)

Curr, J., *The Coal Viewer and Engine Builders Practical Companion* (1797, reprinted with an Introduction by Charles Lee, 1970)

Defoe, Daniel, *A Tour Through the Whole Island of Great Britain* (1724-7)

Dodd, A.H., *The Industrial Revolution in North Wales* (Cardiff, 1951)

Down, C.G. and Warrington, A.J., *The History of the Somerset Coalfield* (Newton Abbot, 1972)

Duckham, B.F., 'Serfdom in Eighteenth Century Scotland', *History*, June 1969
    A History of the Scottish Coal Industry, 1705-1815 (Newton Abbot, 1970)
    Great Pit Disasters (Newton Abbot, 1973)

Engels, F., *The Condition of the Working Class in England in 1844* (1845, reprinted 1969)

Evelyn, J., *Fumifugium: or the Smoke of London Dissipated* (1661)

Ezra, Sir Derek, 'Coal, the Fuel with Reserves', *Coal and Energy Quarterly*, No 1, Summer 1974, pp8-12

Farey, J., Sen, *General View of the Agriculture and Minerals of Derbyshire* (3 vols, 1811-17)

Farey, J., Jun, *A Treatise on the Steam Engine*, Vol 1, (1827, reprinted Newton Abbot, 1971)

Foot, R., *A Plan for Coal* (1945)

Forster, E., *The Keelmen* (Newcastle, 1970)

Gale, W.K.V., *The British Iron & Steel Industry* (Newton Abbot, 1967)

Galloway, R.L., *Annals of Coal Mining and the Coal Trade* (2 vols, 1898, 1904, reprinted Newton Abbot, 1971)

Goodchild, J., *The Industrial and Social Development of South Yorkshire* (Cusworth Hall Museum publication, 1969)
    'The First Mines Inspector in Yorkshire', *South Yorkshire Journal,* Pt 3, (1971), pp15-17
    'On the Introduction of Steam Power into the West Riding', *South Yorkshire Journal,* Pt 3, (1971), pp6-14
Gosden, P.H.J.H., *Self Help* (1973)
Gray, W., *Chorographia, or a Survey of Newcastle-upon-Tyne,* (1649, reprinted Newcastle, 1970)
Greenwell, G.S., *A Glossary of Terms used in the Coal Trade of Northumberland and Durham* (3rd edn, 1888)
Griffin, A.R., *The Miners of Nottinghamshire 1880-1914* (1956)
    The Miners of Nottinghamshire 1914-1944 (1962)
    'Contract Rules in the Notts & Derbys Coalfield', *Bulletin of the Society for the Study of Labour History,* No 16, Spring 1968, pp12-18
    'Bell Pits and Soughs: Some East Midlands Examples', *Industrial Archaeology,* 6, No 4, (1969)
    'Methodism and Trade Unionism in the Nottinghamshire-Derbyshire Coalfield', and 'Primitive Methodism and the Miners' Unions', *Proceedings of the Wesley Historical Society,* February 1969, pp2-9 and October 1969, pp91-2
    'Checkweighing Arrangements at the Butterley Company's Collieries, 1871-3', *Bulletin of the Society for the Study of Labour History,* Spring 1969
    'Industrial Relations by Poster', *Derbys Miscellany,* 5, Pt 2, pp55-68
    'A Boom Colliery of the Boer War Period', *Derbyshire Miscellany,* 5, Pt 4, Autumn 1970, pp213-23 and 6, Pt 3, Spring 1972 pp83-5
    Coalmining (1971)
    Mining in the East Midlands 1550-1947 (1971)
    'Thomas North, Mining Entrepreneur Extraordinary', *Trans of Thoroton Society,* 1972
    'Consultation and Conciliation in the Mining Industry: the need for a New Approach', *Industrial Relations Journal,* Autumn 1972
    'The Miners' Lockout of 1893: A Rejoinder', *Bulletin of the Society for the Study of Labour History,* No 25, Autumn 1972, pp58-65
    'Industrial Archaeology as an aid to the Study of Mining History' *Industrial Archaeology,* February 1974, pp11-28
Griffin, A.R. and Griffin, C.P., 'The Role of Coal Owners' Associations in the East Midlands in the Nineteenth Century', *Renaissance & Modern Studies,* 1973
Griffin, C.P., 'The Economic and Social Development of the Leicestershire & South Derbyshire Coalfield' (unpublished PhD Thesis, Nottingham, 1969)
    'Chartism and the miners in the early 1840s: a critical note' and 'Chartism and the miners' strike of 1842: a reply', *Bulletin of the Society for the Study of Labour History,* No 22, 1971, p21 and No 25, 1972, pp66-8
Hair, T.H., *Sketches of the Coalmines of Northumberland & Durham* (1844, reprinted Newton Abbot, 1969) (with an introduction by Ross, M.)
Hallam, W., *Miners' Leaders* (1894)
Hassall, E.R. and Trickett, J.P., 'The Duke of Bridgewater's Underground Canals', *Trans Manchester Geological & Mining Society,* 15 November 1962
Heyes, D.J., 'The Role of a late Nineteenth Century Colliery Manager' (unpublished

BA Dissertation, Nottingham, 1972)

Holland, J., *The History and Description of Fossil Fuel* (1835)

Hopkinson, G.G., 'Inland Navigations of the Nottinghamshire-Derbyshire Coalfield', *Journal of the Derbys Arch Soc*, 1959

Horne, L.N., 'Benjamin Horne 1719 and Sir John Charrington 1969 are linked', *Charrilock* (House magazine of Charrington, Gardner, Locket & Co Ltd), Winter 1969-70, pp10, 18

J.C., *The Compleat Collier: Or, the whole Art of Sinking, getting and Working Coalmines, etc* (1708, reprinted Newcastle, 1968)

Jevons, H.S., *The British Coal Trade* (1915)

John, A.H., The Industrial Development of South Wales (Cardiff, 1950)

Labour Research Dept, *The Coal Crisis: Facts from the Samuel Commission* (1925-6)

Leifchild, J.R., *Our Coal and Our Coal Pits* (1856)

Lewis, M.J.T., *Early Wooden Railways* (1970)

Liberal Party, *Coal and Power* (1925)

McCutcheon, J., *The Hartley Colliery Disaster* (Seaham, 1962)

MacFarlane, J., 'Denaby Main – a South Yorkshire Mining Village', *Bulletin of the Society for the Study of Labour History*, Autumn 1972, pp82-100

Machen, F., *The Yorkshire Miners* (Barnsley, 1958)

Martin, E.A., *A Piece of Coal* (1896)

Mason, E., *Practical Mining for Coalminers* (1950)

Mayhew, H., 'The Coal Heavers', in Quennell, P. (ed), *Mayhew's London* (1969 edn)

Mee, L.G., 'The Earls Fitzwilliam and the Management of the Collieries and other Industrial Enterprises on the Wentworth Estates' (unpublished PhD Thesis, Nottingham, 1972)

       Aristocratic Enterprise (Glasgow, 1975)

Minchinton, W.E., (ed), *Industrial South Wales 1750-1914* (1969)

Mitchell, B.E. and Deane, P., *Abstract of British Historical Statistics* (Cambridge, 1971)

Mitchell, W., *The Question Answered: 'What Do The Pitmen Want?'* (Bishopwearmouth, 1844)

Moss, K.N., *et al, Historical Review of Coal Mining* (1924)

Mottram, R.H. and Coote, C., *Through Five Generations* (1950)

Nail, M., *The Coal Duties of the City of London and Their Boundary Marks* (1972)

National Coal Board, *Reports of the working Party on Coal Dust Explosions* (1960, 1961)

       Report on Stone Dust Barriers on Coal Conveyor Roads (1961)

       Creswell Model Village (1974)

       Plan for Coal (1974)

National Coal Board, *et al, Coal and Energy Policy in Europe – a report by the British Coal Industry* (1972)

Nef, J.U., *The Rise of the British Coal Industry* (2 vols, 1932)

Neuman, A.M., *Economic Organisation of the British Coal Industry* (1934)

Nixon, F., 'The Early Steam Engine in Derbyshire', *Trans Newcomen Society*, 1957-8, 1958-9

North of England Institute of Mining & Mechanical Engineers, *Centenary Brochure, 1852-1952* (1952)

Parkinson, G., *True Stories of Durham Pit Life* (1912)

Payne, P.L. (ed), *Studies in Scottish Business History* (1967)

Phelps-Brown, E.H., *The Growth of British Industrial Relations* (1959)

Phillipps, E., *The History of the Pioneers of the Welsh Coalfield* (Cardiff, 1925)

Pollard, S., *The Genesis of Modern Management* (1965)

Raybould, T.J., 'The Development and Organisation of Lord Dudley's Mineral Estates 1774-1845', *Econ Hist Rev*, 2nd ser, **XXI**, No 3, 1968, pp529-38

Redmayne, R.A.S., *The British Coal-Mining Industry during the War* (Oxford, 1923)

Rees, Abraham, *New Cyclopaedia* (45 vols, 1802-19), vol 6 Article on Canals (reprinted Newton Abbot, 1974)

Renshaw, P., *The General Strike* (1975)

Riden, P.J., *The Butterley Company 1790-1830* (Chesterfield, 1973)

'The Butterley Company and Railway Construction', *Transport History*, March 1973

Ripley, D., *The Little Eaton Gangway* (Lingfield, 1973)

Robinson, C., *The Energy Crisis and British Coal*, Hobart Paper 59 (Institute of Economic Affairs, 1974)

Rolt, L.T.C., *George and Robert Stephenson* (1960)

Rowe, J.W.F., *Wages in the Coal Industry* (1922)

Sheard, R.L. and Hurst, K.G., 'A History of Water Problems in the South Lancashire Coalfield', *Mining Engineer*, August/September 1973

Slaven, A., 'Earnings and Productivity in the Scottish Coal-mining Industry during the Nineteenth Century: The Dixon Enterprises', *Studies in Scottish Business History* (1967)

Smith, E., *A Pitman's Notebook 1749-51* (ed T. Robertson, Newcastle, 1970)

Smith, R.S., 'The Willoughbys of Wollaton' (unpublished PhD Thesis, Nottingham, 1964)

'Huntingdon Beaumont, Adventurer in Coal Mines', *Renaissance & Modern Studies*, 1, 1956, pp115-53

Storm-Clarke, C., 'The Miners 1870-1970: a Test Case for Oral History', *Victorian Studies*, September 1971

Stuart Smith, C.F., 'On the Winning and Working of Cinderhill Colliery', *Trans North of Eng Inst of Mining & Mech Engrs*, 1861

Swinnerton, H.H., *The Earth Beneath Us* (Harmondsworth, 1955)

Tawney, R.H., 'The problem of the Coal Industry', *The Encyclopaedia of the Labour Movement*, 1927

Taylor, A.J., 'Combination in the Mid-Nineteenth Century Coal Trade, *Trans Royal Hist Soc*, 5th ser, (1953) Vol III, pp23-39

The Sub-Contract System in the British Coal Industry', *Studies in the Industrial Revolution* (1960), pp215-35

Taylor, T.J., *The Archaeology of the Coal Trade* (Newcastle 1858)

Unofficial Reform Committee, *The Miners' Next Step*, (Tonypandy, 1912)

Varley, E., 'Problems of a dear energy economy', *Coal and Energy Quarterly*, No 1, Summer 1974, pp2-7

Victoria History of the County of Nottingham (VCH Notts)

Victoria History of the County of Derby (VCH Derbys)

Ward, J.T. and Wilson, R.G., *Land and Industry* (Newton Abbot, 1971)

Webb, S. and B., *Trade Unionism 1666-1920* (1920)

White, A.W.A., 'The Condition of Mining Labour on a Warwickshire Estate Before

the Industrial Revolution', *Trans Birmingham and Warwickshire Arch Soc*, 84, 1971
Williams, J.E., *The Derbyshire Miners* (1962)
  'Labour in the Coalfields: a critical bibliography' and 'The Miners Lockout of 1893: a rejoinder', *Bulletin of the Society for the Study of Labour History*, No 5, 1962, pp26-7 and No 25, 1972, pp58-65
Wilson, A. and Levy, H., *Workmen's Compensation* (2 vols, 1939)

Index